# ALGEBRA

## For Anyone Who Swears They Suck at Math

### Answer Book

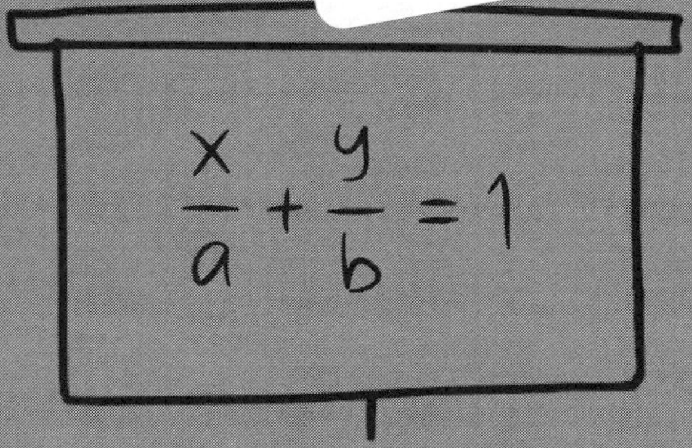


**You Won't Feel Dumb, Stupid or Lost Ever Again**

**Everyone gets stuck.**

Everyone feels a bit dumb at times.

The smart kids.

The teachers.

Anyone.

It's totally normal.

Maths does that to you.

But what matters is what you do when you get stuck.

You don't wanna get frustrated and give up.

Here's what to do.

First.

Stop and think.

Why are you stuck?

Give it 5 minutes max.

Don't dwell on it.

Is there something you don't get?

Could there be a silly mistake?

Or is it just you have no idea where to start?

What is it, exactly?

Go back and have a look at the examples.

Carefully follow each step.

See if you can figure out what you're doing wrong.

Or what you're not quite getting.

If you're still stuck after 5 minutes.

Save yourself the headache.

Check the step-by-step answers.

See how the question's answered.

Understand it.

Then go back and try it again.

If you get it right.

Great.

Move on to the next question.

If not.

Look at the steps again.

Pin  point your mistake.

Where are you going wrong?

Learn from it.

Then try the question again.

Until you get it right.

Now, if you're checking the answer book a bit too much.

Chances are there's something you're not getting.

Maybe it’s something from the topic you’re on.

Maybe something from earlier.

Either way, go back to where things last made sense.

And build back up from there.

Find what you’re not getting.

Fix it.

Then move forward again.

## Stuff You're About to Learn How to Answer

## Linear Equations



## Simultaneous Equations

The Basics

# Divisibility Rules

## Multiple Choice

1 **Option A**

$7 + 1 + 4 = 9$

$9$ is divisible by $3$.

$\therefore$ $714$ is divisible by $3$.

2 **Option C**

Divisible by 10: Last digit must be 0.

3 **Option D**

528 is even.

$5 + 2 + 8 = 15$

$15$ is divisible by $3$.

$\therefore$ $528$ is divisible by 6.

4 **Option B**

$4 + 6 + 8 = 18$

$18$ is divisible by $9$.

$\therefore$ $468$ is divisible by $9$.

Missing digit is 6

5 **Option D**

The last two digits are 36.

36 is divisible by $4$.

$\therefore$ Any first digit will make number divisible by $4$.

## Time To Practice

**1** $24$

Last digit is even

→ 4 divisible by 2

∴ Yes

**2** $42$

Sum of digits $4 + 2 = 6$

→ 6 divisible by 3

∴ Yes

**3** $45$

Last digit is 5

→ 5 divisible by 5

∴ Yes

**4** $68$

Last digit is even.

Sum of digits $6 + 8 = 14$

→ 14 NOT divisible by 3

∴ No

**5** $72$

Using normal division $72 \div 8 = 9$

→ 72 divisible by 8

∴ Yes

**6** $84$

Last digit is even

Sum of digits $8 + 4 = 12$

→ 12 is divisible by 3

∴ Yes

**7** $84$

$8 - (2 \times 4) = 0$

→ 0

∴ Yes

**8** $94$

$9 - (2 \times 4) = 1$

→ 1 NOT divisible by 7

∴ No

**9** $98$

Last two digit 98

→ 98 is NOT divisible by 4

∴ No

**10** $105$

Sum of digits $1 + 0 + 5 = 6$

→ 6 is NOT divisible by 9

∴ No

**11** $114$

$11 - (2 \times 4) = 3$

$\rightarrow$ 3 NOT divisible by 7

$\therefore$ No

**12** $144$

Last two digit 44

$\rightarrow$ 44 is divisible by 4

$\therefore$ Yes

**13** $120$

**a** **2:** Last digit is even $\rightarrow$ divisible by 2

**3:** Sum of digits $1 + 2 + 3 = 68$ $\rightarrow$ divisible by 3

**4:** Last 2 is digits $20$ $(20 \div 4 = 5)$ $\rightarrow$ is divisible by 4

**5:** Last digit is 0 $\rightarrow$ divisible by 5

**6:** Even and divisible by 3 $\rightarrow$ divisible by 6

**7:** $12 - (2 \times 0) = 12 \rightarrow$ NOT divisible by 7

**8:** Last three digits $120 \div 8 = 15$ $\rightarrow$ divisible by 8

**9:** Sum of digits $1 + 2 + 3 = 6$ $\rightarrow$ NOT divisible by 9

**10:** Last digit is 0 $\rightarrow$ divisible by 10

Answer: $2, 3, 4, 5, 6, 8, 10$

**b** $156$

**2:** Last digit is even $\rightarrow$ divisible by 2

**3:** Sum of digits $1 + 5 + 6 = 12$ $\rightarrow$ divisible by 3

**4:** Last 2 is digits $56$ $(56 \div 4 = 14)$ $\rightarrow$ is divisible by 4

**5:** Last digit is 6 $\rightarrow$ NOT divisible by 5

**6:** Even and divisible by 3 $\rightarrow$ divisible by 6

**7:** $15 - (2 \times 6) = 3 \rightarrow$ NOT divisible by 7

**8:** Last three digits $120 \div 8 = 19.5$ $\rightarrow$ NOT divisible by 8

**9:** Sum of digits $1 + 5 + 6 = 12$ $\rightarrow$ NOT divisible by 9

**10:** Last digit is 6 $\rightarrow$ NOT divisible by 10

Answer: $2, 3, 4, 6$

**c** $198$

**2:** Last digit is even → divisible by 2

**3:** Sum of digits $1 + 9 + 8 = 18$ → divisible by 3

**4:** Last 2 is digits $98$ $(98 \div 4 = 24.5)$ → NOT divisible by 4

**5:** Last digit is 8 → NOT divisible by 5

**6:** Even and divisible by 3 → divisible by 6

**7:** $19 - (2 \times 8) = 3$ → NOT divisible by 7

**8:** Last three digits $198 \div 8 = 24.75$ → NOT divisible by 8

**9:** Sum of digits $1 + 9 + 8 = 18$ → divisible by 9

**10:** Last digit is 8 → NOT divisible by 10

Answer: $2, 3, 6, 9$

**d** $240$

**2, 5 & 10:** Last digit is 0→ divisible by 2, 5 & 10

**3:** Sum of digits $2 + 4 + 0 = 6$ → divisible by 3

**4:** Last 2 is digits $40$ $(40 \div 4 = 10)$ → is divisible by 4

**6:** Even and divisible by 3 → divisible by 6

**7:** $24 - (2 \times 0) = 24$ → NOT divisible by 7

**8:** Last three digits $120 \div 8 = 30$ → divisible by 8

**9:** Sum of digits $2 + 4 + 0 = 6$ → NOT divisible by 9

Answer: $2, 3, 4, 5, 6, 8, 10$

**e** $357$

**2:** Last digit is odd → NOT divisible by 2

**3:** Sum of digits $3 + 5 + 7 = 15$ → divisible by 3

**4:** Last 2 is digits $57$ $(57 \div 4 = 14.25)$ → NOT divisible by 4

**5 & 10:** Last digit is 7 → NOT divisible by 5 or 10

**6:** Last digit is odd → NOT divisible by 6

**7:** $35 - (2 \times 7) = 21$ → divisible by 7

**8:** Last three digits $357 \div 8 = 44.625$ → NOT divisible by 8

**9:** Sum of digits $3 + 5 + 7 = 15$ → NOT divisible by 9

Answer: $3, 7$

**f** $378$

**2:** Last digit is even → divisible by 2

**3:** Sum of digits $3 + 7 + 8 = 18$ → divisible by 3

**4:** Last 2 is digits $78$ $(78 \div 4 = 19.5)$ → NOT divisible by 4

**5:** Last digit is 8 → NOT divisible by 5

**6:** Even and divisible by 3 → divisible by 6

**7:** $37 - (2 \times 8) = 21$ → divisible by 7

**8:** Last three digits $378 \div 8 = 47.25$ → NOT divisible by 8

**9:** Sum of digits $3 + 7 + 8 = 18$ → divisible by 9

**10:** Last digit is 8 → NOT divisible by 10

Answer: $2, 3, 6, 7, 9$

**g** $434$

**2:** Last digit is even → divisible by 2

**3:** Sum of digits $4 + 3 + 4 = 11$ → NOT divisible by 3

**4:** Last 2 is digits $34$ $(34 \div 4 = 8.5)$ → NOT divisible by 4

**5:** Last digit is 8 → NOT divisible by 5

**6:** Not divisible by 3 → NOT divisible by 6

**7:** $43 - (2 \times 4) = 35$ → divisible by 7

**8:** Last three digits $378 \div 8 = 54.25$ → NOT divisible by 8

**9:** Sum of digits $4 + 3 + 4 = 11$ → NOT divisible by 9

**10:** Last digit is 4 → NOT divisible by 10

Answer: $2, 7$

**h** $480$

**2:** Last digit is even → divisible by 2

**3:** Sum of digits $4 + 8 + 0 = 12$ → divisible by 3

**4:** Last 2 is digits $80$ $(80 \div 4 = 20)$ → divisible by 4

**5:** Last digit is 0 → divisible by 5

**6:** Even and divisible by 3 → divisible by 6

**7:** $48 - (2 \times 0) = 48$ → NOT divisible by 7

**8:** Last three digits $480 \div 8 = 60$ → divisible by 8

**9:** Sum of digits $4 + 8 + 0 = 12$ → NOT divisible by 9

**10:** Last digit is 0 → divisible by 10

Answer: $2, 3, 4, 5, 6, 8, 10$

i $496$

**2:** Last digit is even → divisible by 2

**3:** Sum of digits $4 + 9 + 6 = 19$ → NOT divisible by 3

**4:** Last 2 is digits $96$ $(96 \div 4 = 24)$ → divisible by 4

**5:** Last digit is 6 → NOT divisible by 5

**6:** Not divisible by 3 → NOT divisible by 6

**7:** $49 - (2 \times 6) = 37 \rightarrow$ NOT divisible by 7

**8:** Last three digits $378 \div 8 = 62$ → divisible by 8

**9:** Sum of digits $4 + 9 + 6 = 19$ → NOT divisible by 9

**10:** Last digit is 6 → NOT divisible by 10

Answer: $2, 4, 8$

j $504$

**2:** Last digit is even → divisible by 2

**3:** Sum of digits $5 + 0 + 4 = 9$ → divisible by 3

**4:** Last 2 is digits $80$ $(4 \div 4 = 1)$ → divisible by 4

**5:** Last digit is 4 → NOT divisible by 5

**6:** Even and divisible by 3 → divisible by 6

**7:** $50 - (2 \times 4) = 42 \rightarrow$ divisible by 7

**8:** Last three digits $504 \div 8 = 63$ → divisible by 8

**9:** Sum of digits $5 + 0 + 4 = 9$ → divisible by 9

**10:** Last digit is 4 → NOT divisible by 10

Answer: $2, 3, 4, 6, 7, 8, 9$

**k** $739$

**2, 4, 5, 6, 8, 10:** Last digit is odd → NOT divisible by 2, 4, 5, 6, 8, 10

**3:** Sum of digits $7 + 3 + 9 = 19$ → NOT divisible by 3

**7:** $73 - (2 \times 9) = 55 \rightarrow$ NOT divisible by 7

**9:** Sum of digits $7 + 3 + 9 = 19$ → NOT divisible by 9

Answer: Not divisible by any number from 2 to 10

**l** $840$

**2, 5 & 10:** Last digit is 0→ divisible by 2, 5 and 10

**3:** Sum of digits $8 + 4 + 0 = 12$ → divisible by 3

**4:** Last 2 is digits $40$ $(40 \div 4 = 10)$ → divisible by 4

**6:** Even and divisible by 3 → divisible by 6

**7:** $84 - (2 \times 0) = 84 \rightarrow$ divisible by 7

**8:** Last three digits $504 \div 8 = 105$ → divisible by 8

**9:** Sum of digits $8 + 4 + 0 = 12$ → NOT divisible by 9

Answer: $2, 3, 4, 5, 6, 7, 8, 10$

## Challenge Yourself

1 Sum of digits must be divisible by 3 but not 9.

$3 + 1 + 8 = 12$

$3 + 4 + 8 = 15$

328, 338, 358, 368, 388 and 398 are NOT divisible by 3.

378 is divisible by 9.

318 and 348 are divisible by 3 but not 9 ∴ the missing digits are 1 or 4.

2 **Divisibility by 4**: The last two digits must be divisible by 4. Possible options are 20, 24, or 28.

**Divisibility by 7:** Check which of these numbers 720, 724, 728 is divisible by 7.

$720: 72 - (2 \times 0) = 72$ → NOT divisible by 7

$724: 72 - (2 \times 4) = 64$ → NOT divisible by 7

$728: 72 - (2 \times 8) = 56$ → divisible by 7

728 is divisible by 4 and 7 ∴ the missing digit is 8.

3 **Divisibility by 2 and 5**: The last digit is 0, so any first digit will make number divisible by 2 and 5.

**Divisibility by 7:** Test each digit from 1-9 to find the first digit that makes the number divisible by 7.

$150:\ 15 - (2 \times 0) = 15$ → NOT divisible by 7

$250:\ 25 - (2 \times 0) = 25$ → divisible by 7

$350:\ 35 - (2 \times 0) = 35$ → divisible by 7

350 is divisible by 2, 5 and 7 ∴ the smallest possible missing digit is 3.

4 **Divisibility by 3 and 9:** The sum of the digits must be divisible by $3$ and $9$.

$6 + 0 + 3 = 9$

$6 + 9 + 3 = 18$

$613, 623, 633, 643, 653, 663, 673$ and $683$ are NOT divisible by both $3$ and $9$.

**Divisibility by 7: Check which of these numbers $603$, or $693$ is divisible by $7$.**

$603{:}\ 60 - (2 \times 3) = 54 \rightarrow$ NOT divisible by 7

$693{:}\ 69 - (2 \times 3) = 63 \rightarrow$ divisible by 7

$693$ is divisible by $3, 9$ and $7$ $\therefore$ the missing digit is $9$.

5 **Divisibility by 6**: The last digit is even so number must be divisible by $3$.
**Divisibility by 3**: The sum of digits must be divisible by $3$.

$2 + 0 + 4 = 6$

$5 + 0 + 4 = 9$

$8 + 0 + 4 = 12$

$204, 504$ and $804$ are divisible by 2 and $3$ $\therefore$ divisible by $6$.

$304, 404, 604, 704, 904$ are NOT divisible by $3$.

**Divisibility by 7**:

$204{:}\ 20 - (2 \times 4) = 12 \rightarrow$ NOT divisible by 7

$504{:}\ 50 - (2 \times 4) = 42 \rightarrow$ divisible by 7

$804{:}\ 80 - (2 \times 4) = 72 \rightarrow$ NOT divisible by 7

$504$ is also divisible by 7 $\therefore$ the missing digit is 5 (as no other option is left)

**Divisibility by 8**: Use long division to double check

$504 \div 8 = 63$

# Simplifying Fractions

## Multiple Choice

1 **Option B**

$$\frac{9}{18}$$

$$= \frac{1 \times 9}{2 \times 9}$$

$$= \frac{1}{2}$$

2 **Option A**

$$\frac{12}{15}$$

$$= \frac{3 \times 4}{3 \times 5}$$

$$= \frac{4}{5}$$

3 **Option C**

$$\frac{6}{54}$$

$$= \frac{6 \times 1}{6 \times 9}$$

$$= \frac{1}{9}$$

4 **Option B**

$$\frac{42}{48}$$

$$= \frac{7 \times 6}{8 \times 6}$$

$$= \frac{7}{8}$$

5 **Option D**

$$\frac{12}{16} = \frac{3 \times 4}{4 \times 4} = \frac{3}{4}$$

$$\frac{21}{28} = \frac{7 \times 3}{7 \times 4} = \frac{3}{4}$$

$$\frac{33}{44} = \frac{11 \times 3}{11 \times 4} = \frac{3}{4}$$

$$\frac{20}{24} = \frac{4 \times 5}{4 \times 6} = \frac{5}{6}$$

## Time To Practice

**1** **a** $\frac{4}{36}$

$= \frac{4 \times 1}{4 \times 9}$

$= \frac{1}{9}$

**b** $\frac{5}{35}$

$= \frac{5 \times 1}{5 \times 7}$

$= \frac{1}{7}$

**c** $\frac{6}{18}$

$= \frac{6 \times 1}{6 \times 3}$

$= \frac{1}{3}$

**2** **a** $\frac{9}{63}$

$= \frac{9 \times 1}{9 \times 7}$

$= \frac{1}{7}$

**b** $\frac{8}{56}$

$= \frac{8 \times 1}{8 \times 8}$

$= \frac{1}{8}$

**c** $\frac{7}{77}$

$= \frac{7 \times 1}{7 \times 11}$

$= \frac{1}{11}$

**3** **a** $\frac{14}{21}$

$= \frac{7 \times 2}{7 \times 3}$

$= \frac{2}{3}$

**b** $\frac{16}{40}$

$= \frac{8 \times 2}{8 \times 5}$

$= \frac{2}{5}$

**c** $\frac{18}{81}$

$= \frac{9 \times 2}{9 \times 9}$

$= \frac{2}{9}$

**4** **a** $\frac{21}{35}$

$= \frac{7 \times 3}{7 \times 5}$

$= \frac{3}{5}$

**b** $\frac{22}{55}$

$= \frac{11 \times 2}{11 \times 5}$

$= \frac{2}{5}$

**c** $\frac{24}{36}$

$= \frac{12 \times 2}{12 \times 3}$

$= \frac{2}{3}$

**5** **a** $\frac{28}{63}$

$= \frac{7 \times 4}{7 \times 9}$

$= \frac{4}{9}$

**b** $\frac{30}{35}$

$= \frac{5 \times 6}{5 \times 7}$

$= \frac{6}{7}$

**c** $\frac{49}{56}$

$= \frac{7 \times 7}{7 \times 8}$

$= \frac{7}{8}$

**6** **a** $\frac{50}{80}$

$= \frac{10 \times 5}{10 \times 8}$

$= \frac{5}{8}$

**b** $\frac{55}{99}$

$= \frac{11 \times 5}{11 \times 9}$

$= \frac{5}{9}$

**c** $\frac{60}{72}$

$= \frac{12 \times 5}{12 \times 6}$

$= \frac{5}{6}$

**7** **a** $\frac{72}{99}$

$= \frac{9 \times 8}{9 \times 11}$

$= \frac{8}{11}$

**b** $\frac{84}{108}$

$= \frac{12 \times 7}{12 \times 9}$

$= \frac{7}{9}$

**c** $\frac{132}{144}$

$= \frac{12 \times 11}{12 \times 12}$

$= \frac{11}{12}$

## Challenge Yourself

**1** **a** $\frac{80}{120}$

$= \frac{10 \times 8}{10 \times 12}$

$= \frac{8}{12} = \frac{4 \times 2}{4 \times 3}$

$= \frac{2}{3}$

**b** $\frac{54}{72}$

$= \frac{9 \times 6}{9 \times 8}$

$= \frac{6}{8} = \frac{2 \times 3}{2 \times 4}$

$= \frac{3}{4}$

**c** $\frac{88}{110}$

$= \frac{11 \times 8}{11 \times 10}$

$= \frac{8}{10} = \frac{2 \times 4}{2 \times 5}$

$= \frac{4}{5}$

**2** **a** $\frac{120}{144}$

$= \frac{12 \times 10}{12 \times 12}$

$= \frac{10}{12} = \frac{2 \times 5}{2 \times 6}$

$= \frac{5}{6}$

**b** $\frac{125}{175}$

$= \frac{5 \times 25}{5 \times 35}$

$= \frac{25}{35} = \frac{5 \times 5}{5 \times 7}$

$= \frac{5}{7}$

**c** $\frac{300}{330}$

$= \frac{10 \times 30}{10 \times 33}$

$= \frac{30}{33} = \frac{3 \times 10}{3 \times 11}$

$= \frac{10}{11}$

# Adding and Subtracting Fractions

## Multiple Choice

**1** **Option D**

$$\frac{4}{5}-\frac{2}{3}$$

$$=\frac{12-10}{15}$$

$$=\frac{2}{15}$$

**2** **Option A**

$$\frac{11}{12}-\frac{5}{6}=\frac{11}{12}-\frac{5\times 2}{6\times 2}$$

$$=\frac{11}{12}-\frac{10}{12}$$

$$=\frac{1}{12}$$

**3** **Option B**

$$\frac{7}{10}-1$$

$$=\frac{7}{10}-\frac{10}{10}$$

$$=-\frac{3}{10}$$

**4** **Option B**

$$1+\left(\frac{3}{5}-\frac{2}{7}\right)=1+\left(\frac{21-10}{35}\right)$$

$$=1+\frac{11}{35}$$

$$=\frac{35}{35}+\frac{11}{35}$$

$$=\frac{46}{35}$$

**5** **Option C**

$$\frac{2}{3}-\left(\frac{5}{6}-\frac{1}{4}\right)=\frac{2}{3}-\left(\frac{20-6}{24}\right)=\frac{2}{3}-\frac{14}{24}$$

$$=\frac{2}{3}-\frac{7}{12}=\frac{2\times 4}{3\times 4}-\frac{7}{12}$$

$$=\frac{8}{12}-\frac{7}{12}$$

$$=\frac{1}{12}$$

## Time To Practice

**1** **a** $\frac{1}{4}+\frac{2}{4}$

$=\frac{3}{4}$

**b** $\frac{3}{8}-\frac{1}{8}$

$=\frac{2}{8}=\frac{1}{4}$

**c** $\frac{1}{10}+\frac{9}{10}$

$=\frac{10}{10}=1$

**2** **a** $\frac{5}{12}-\frac{1}{4}$

$=\frac{5}{12}-\frac{3}{12}$

$=\frac{2}{12}=\frac{1}{6}$

**b** $-\frac{7}{10}+\frac{4}{5}$

$=-\frac{7}{10}+\frac{8}{10}$

$=\frac{1}{10}$

**c** $\frac{2}{3}-\frac{5}{6}$

$=\frac{4}{6}-\frac{5}{6}$

$=-\frac{1}{6}$

**3** **a** $\frac{1}{8}+\frac{3}{4}$

$=\frac{4+24}{32}$

$=\frac{28}{32}=\frac{4\times 7}{4\times 8}$

$=\frac{7}{8}$

**b** $\frac{4}{7}+\frac{3}{10}$

$=\frac{40+21}{70}$

$=\frac{61}{70}$

**c** $\frac{9}{11}-\frac{3}{8}$

$=\frac{72-33}{88}$

$=\frac{39}{88}$

**4** **a** $2-\frac{1}{4}$

$=\frac{2}{1}-\frac{1}{4}$

$=\frac{8-1}{4}$

$=\frac{7}{4}$

**b** $-1+\frac{1}{6}$

$=-\frac{1}{1}+\frac{1}{6}$

$=\frac{-6+1}{6}$

$=-\frac{5}{6}$

**c** $3+\frac{7}{10}$

$=\frac{3}{1}+\frac{7}{10}$

$=\frac{30+7}{10}$

$=\frac{37}{10}$

**5** **a**

$$1+\left(\frac{7}{24}-\frac{5}{12}\right)$$

$$=1+\left(\frac{7}{24}-\frac{10}{24}\right)$$

$$=1+\left(-\frac{3}{24}\right)=1-\frac{1}{8}$$

$$=\frac{8}{8}-\frac{1}{8}$$

$$=\frac{7}{8}$$

**b**

$$-1+\left(2-\frac{1}{2}\right)$$

$$=-1+\left(\frac{2}{1}-\frac{1}{2}\right)$$

$$=-1+\left(\frac{4-1}{2}\right)=-1+\frac{3}{2}$$

$$=-\frac{2}{2}+\frac{3}{2}$$

$$=\frac{1}{2}$$

**c**

$$3-\left(\frac{9}{10}+\frac{2}{5}\right)$$

$$=3-\left(\frac{9}{10}+\frac{4}{10}\right)$$

$$=3-\left(\frac{13}{10}\right)=3-\frac{13}{10}$$

$$=\frac{3}{1}-\frac{13}{10}$$

$$=\frac{30-13}{10}$$

$$=\frac{17}{10}$$

**6** **a** $\left(\frac{1}{4}+\frac{2}{5}\right)-\frac{1}{10}$

$=\left(\frac{5+8}{20}\right)-\frac{1}{10}$

$=\frac{13}{20}-\frac{1}{10}$

$=\frac{13}{20}-\frac{2}{20}$

$=\frac{11}{20}$

**b** $-\frac{4}{9}+\left(\frac{1}{3}-\frac{1}{12}\right)$

$=-\frac{4}{9}+\left(\frac{4}{12}-\frac{1}{12}\right)$

$=-\frac{4}{9}+\frac{3}{12}$

$=\frac{-48+27}{108}$

$=-\frac{21}{108}=-\frac{3\times 7}{3\times 36}$

$=-\frac{7}{36}$

**c** $\left(-\frac{1}{6}+\frac{1}{4}\right)+\frac{5}{8}$

$=\left(\frac{-4+6}{24}\right)+\frac{5}{8}$

$=\frac{2}{24}+\frac{5}{8}$

$=\frac{2}{24}+\frac{5}{8}$

$=\frac{2}{24}+\frac{15}{24}$

$=\frac{17}{24}$

**7** **a**

$$\left(-\frac{7}{10}-\frac{2}{3}\right)-\frac{1}{5}$$

$$=\left(\frac{-21-20}{30}\right)-\frac{1}{5}$$

$$=-\frac{41}{30}-\frac{1}{5}$$

$$=-\frac{41}{30}-\frac{6}{30}$$

$$=-\frac{47}{30}$$

**b**

$$\frac{3}{5}+\left(-\frac{3}{4}-\frac{1}{10}\right)$$

$$=\frac{3}{5}+\left(\frac{-30-4}{40}\right)$$

$$=\frac{3}{5}+\left(-\frac{34}{40}\right)$$

$$=\frac{3}{5}-\frac{34}{40}$$

$$=\frac{3}{5}-\frac{17}{20}$$

$$=\frac{12}{20}-\frac{17}{20}$$

$$=-\frac{5}{20}$$

$$=-\frac{1}{4}$$

**c**

$$\left(-\frac{2}{3}-\frac{4}{5}\right)+\frac{2}{3}$$

$$=\left(\frac{-10-12}{15}\right)+\frac{2}{3}$$

$$=-\frac{22}{15}+\frac{2}{3}$$

$$=-\frac{22}{15}+\frac{10}{15}$$

$$=-\frac{12}{15}=-\frac{3\times 4}{3\times 5}$$

$$=-\frac{4}{5}$$

## Challenge Yourself

**1** **a** $\left(\frac{8}{13}+\frac{1}{2}\right)-\frac{3}{26}$

$=\left(\frac{16+13}{26}\right)-\frac{3}{26}$

$=\frac{29}{26}-\frac{3}{26}$

$=\frac{26}{26}$

$=1$

**b** $-\frac{11}{20}+\left(2-\frac{1}{2}\right)$

$=-\frac{11}{20}+\left(\frac{2}{1}-\frac{1}{2}\right)$

$=-\frac{11}{20}+\left(\frac{4-1}{2}\right)$

$=-\frac{11}{20}+\frac{3}{2}$

$=-\frac{11}{20}+\frac{30}{20}$

$=\frac{19}{20}$

**2** **a** $\frac{1}{3}-\left(\frac{5}{6}-\frac{7}{18}\right)$

$=\frac{1}{3}-\left(\frac{15}{18}-\frac{7}{18}\right)$

$=\frac{1}{3}-\frac{8}{18}$

$=\frac{1}{3}-\frac{4}{9}$

$=\frac{3}{9}-\frac{4}{9}$

$=-\frac{1}{9}$

**b** $-\frac{2}{5}-\left(\frac{1}{10}-\frac{1}{15}\right)$

$=-\frac{2}{5}-\left(\frac{15-10}{150}\right)$

$=-\frac{2}{5}-\frac{5}{150}$

$=-\frac{2}{5}-\frac{1}{30}$

$=-\frac{12}{30}-\frac{1}{30}$

$=-\frac{13}{30}$

**3** **a** $\frac{2}{18}-\left(\frac{2}{9}-\frac{4}{27}\right)$

$=\frac{1}{9}-\left(\frac{6}{27}-\frac{4}{27}\right)$

$=\frac{1}{9}-\frac{2}{27}$

$=\frac{3}{27}-\frac{2}{27}$

$=\frac{1}{27}$

**b** $\left(\frac{5}{8}-\frac{1}{10}\right)-\frac{3}{4}$

$=\left(\frac{50-8}{80}\right)-\frac{3}{4}$

$=\frac{42}{80}-\frac{3}{4}$

$=\frac{21}{40}-\frac{3}{4}$

$=\frac{21}{40}-\frac{30}{40}$

$=-\frac{9}{40}$

# Multiplying and Dividing Fractions

## Multiple Choice

**1** **Option B**

$$\frac{2}{5} \times \frac{3}{4}$$

$$= \frac{6}{20}$$

$$= \frac{3}{10}$$

**2** **Option D**

$$\frac{5}{6} \div \frac{10}{9} = \frac{5}{6} \times \frac{9}{10}$$

$$= \frac{1}{2} \times \frac{3}{2}$$

$$= \frac{3}{4}$$

**3** **Option C**

$$6 \div \frac{2}{5} = 6 \times \frac{5}{2}$$

$$= \frac{30}{2}$$

$$= 15$$

**4** **Option B**

$$\frac{9}{14} \times \frac{7}{3}$$

$$= \frac{3}{2} \times \frac{1}{1}$$

$$= \frac{3}{2}$$

**5** **Option A**

$$\frac{5}{8} \div 4 = \frac{5}{8} \div \frac{4}{1}$$

$$= \frac{5}{8} \times \frac{1}{4}$$

$$= \frac{5}{32}$$

## Time To Practice

**1** **a** $\frac{2}{9} \times \frac{1}{5}$

$= \frac{2}{45}$

**b** $\frac{1}{5} \times \frac{3}{2}$

$= \frac{3}{10}$

**c** $\frac{7}{8} \times \frac{9}{11}$

$= \frac{63}{88}$

**2** **a** $\frac{2}{3} \times 4$

$= \frac{8}{3}$

**b** $\frac{2}{3} \div 4$

$= \frac{2}{3} \times \frac{1}{4}$

$= \frac{2}{12} = \frac{1}{6}$

**c** $4 \div \frac{2}{3}$

$= 4 \times \frac{3}{2}$

$= \frac{12}{2} = 6$

**3** **a** $\frac{5}{6} \times \frac{18}{25}$

$= \frac{1}{1} \times \frac{3}{5}$

$= \frac{3}{5}$

**b** $-\frac{7}{9} \times \frac{45}{49}$

$= -\frac{1}{1} \times \frac{5}{7}$

$= -\frac{5}{7}$

**c** $\frac{3}{40} \times \frac{5}{33}$

$= \frac{1}{8} \times \frac{1}{11}$

$= \frac{1}{88}$

**4** **a** $\frac{1}{6} \div \frac{4}{5}$

$= \frac{1}{6} \times \frac{5}{4}$

$= \frac{5}{24}$

**b** $\frac{2}{3} \div \frac{7}{8}$

$= \frac{2}{3} \times \frac{8}{7}$

$= \frac{16}{21}$

**c** $\frac{4}{9} \div \frac{5}{8}$

$= \frac{4}{9} \times \frac{8}{5}$

$= \frac{32}{45}$

**5** **a** $\left(\frac{2}{3} \div \frac{1}{4}\right) \div 2$

$= \left(\frac{2}{3} \times \frac{4}{1}\right) \div 2$

$= \frac{8}{3} \times \frac{1}{2}$

$= \frac{8}{6}$

$= \frac{4}{3}$

**b** $\frac{5}{9} \div \left(\frac{5}{6} \div \frac{2}{3}\right)$

$= \frac{5}{9} \div \left(\frac{5}{6} \times \frac{3}{2}\right)$

$= \frac{5}{9} \div \left(\frac{5}{2} \times \frac{1}{2}\right)$

$= \frac{5}{9} \div \frac{5}{4}$

$= \frac{5}{9} \times \frac{4}{5}$

$= \frac{4}{9}$

**c** $\left(\frac{1}{2} \div \frac{45}{40}\right) \div \frac{40}{25}$

$= \left(\frac{1}{2} \div \frac{9}{8}\right) \div \frac{8}{5}$

$= \left(\frac{1}{2} \times \frac{8}{9}\right) \div \frac{8}{5}$

$= \left(\frac{1}{1} \times \frac{4}{9}\right) \div \frac{8}{5}$

$= \frac{4}{9} \times \frac{5}{8}$

$= \frac{1}{9} \times \frac{5}{2}$

$= \frac{5}{18}$

**6** **a** $\left(\frac{1}{2} - \frac{2}{3}\right) \times \frac{3}{4}$

$= \left(\frac{3-4}{6}\right) \times \frac{3}{4}$

$= -\frac{1}{6} \times \frac{3}{4}$

$= -\frac{1}{2} \times \frac{1}{4}$

$= -\frac{1}{8}$

**b** $\frac{3}{5} \div \left(\frac{1}{4} + \frac{1}{6}\right)$

$= \frac{3}{5} \div \left(\frac{6+4}{24}\right)$

$= \frac{3}{5} \div \frac{10}{24}$

$= \frac{3}{5} \div \frac{5}{12}$

$= \frac{3}{5} \times \frac{12}{5}$

$= \frac{36}{25}$

**c** $\left(\frac{5}{8} \times \frac{2}{3}\right) - \frac{1}{4}$

$= \left(\frac{5}{4} \times \frac{1}{3}\right) - \frac{1}{4}$

$= \frac{5}{12} - \frac{1}{4}$

$= \frac{5}{12} - \frac{3}{12}$

$= \frac{2}{12}$

$= \frac{1}{6}$

**7** **a** $\left(\frac{12}{25} \div \frac{8}{5}\right) \times \frac{7}{10}$

$= \left(\frac{12}{25} \times \frac{5}{8}\right) \times \frac{7}{10}$

$= \left(\frac{3}{5} \times \frac{1}{2}\right) \times \frac{7}{10}$

$= \frac{3}{10} \times \frac{7}{10}$

$= \frac{21}{100}$

**b** $\left(\frac{55}{30} \div \frac{40}{30}\right) \times \frac{10}{11}$

$= \left(\frac{11}{6} \div \frac{4}{3}\right) \times \frac{10}{11}$

$= \left(\frac{11}{6} \times \frac{3}{4}\right) \times \frac{10}{11}$

$= \frac{33}{24} \times \frac{10}{11}$

$= \frac{3}{12} \times \frac{5}{1}$

$= \frac{1}{4} \times \frac{5}{1}$

$= \frac{5}{4}$

**c** $\frac{5}{2} \times \left(\frac{160}{30} \div \frac{2}{15}\right)$

$= \frac{5}{2} \times \left(\frac{16}{3} \times \frac{15}{2}\right)$

$= \frac{5}{2} \times \left(\frac{8}{1} \times \frac{5}{1}\right)$

$= \frac{5}{2} \times \frac{40}{1}$

$= \frac{5}{1} \times \frac{20}{1}$

$= 100$

## Challenge Yourself

**1** **a** $\left(\frac{8}{9}\div\frac{2}{3}\right)\div\left(\frac{3}{4}\div\frac{1}{2}\right)$

$=\left(\frac{8}{9}\times\frac{3}{2}\right)\div\left(\frac{3}{4}\times\frac{2}{1}\right)$

$=\left(\frac{4}{3}\times\frac{1}{1}\right)\div\left(\frac{3}{2}\times\frac{1}{1}\right)$

$=\frac{4}{3}\div\frac{3}{2}$

$=\frac{4}{3}\times\frac{2}{3}$

$=\frac{8}{9}$

**b** $\left(\frac{11}{12}\div\frac{5}{6}\right)\div\left(\frac{3}{14}\div\frac{1}{4}\right)$

$=\left(\frac{11}{12}\times\frac{6}{5}\right)\div\left(\frac{3}{14}\times\frac{4}{1}\right)$

$=\left(\frac{11}{2}\times\frac{1}{5}\right)\div\left(\frac{3}{7}\times\frac{2}{1}\right)$

$=\frac{11}{10}\div\frac{6}{7}$

$=\frac{11}{10}\times\frac{7}{6}$

$=\frac{77}{60}$

**2** **a** $\left(\frac{5}{8}\div\frac{1}{4}\right)\times\left(\frac{3}{5}-\frac{1}{10}\right)$

$=\left(\frac{5}{8}\times\frac{4}{1}\right)\times\left(\frac{30-5}{50}\right)$

$=\left(\frac{5}{2}\times\frac{1}{1}\right)\times\left(\frac{25}{50}\right)$

$=\frac{5}{2}\times\frac{1}{2}$

$=\frac{5}{4}$

**b** $\frac{4}{5}\times\left(\frac{2}{3}+\frac{1}{6}\right)\div\frac{3}{4}$

$=\frac{4}{5}\times\left(\frac{4}{6}+\frac{1}{6}\right)\div\frac{3}{4}$

$=\frac{4}{5}\times\frac{5}{6}\div\frac{3}{4}$

$=\frac{2}{1}\times\frac{1}{3}\div\frac{3}{4}$

$=\frac{2}{3}\times\frac{4}{3}$

$=\frac{8}{9}$

3 a

$$\left(\frac{18}{25} \div \frac{12}{35}\right) + \left(\frac{16}{45} \times \frac{27}{32}\right)$$

$$= \left(\frac{18}{25} \times \frac{35}{12}\right) + \left(\frac{1}{5} \times \frac{3}{2}\right)$$

$$= \left(\frac{3}{5} \times \frac{7}{2}\right) + \frac{3}{10}$$

$$= \frac{21}{10} + \frac{3}{10}$$

$$= \frac{24}{10}$$

$$= \frac{12}{5}$$

b

$$\left(\frac{24}{35} \times \frac{15}{28}\right) - \left(\frac{3}{14} \div \frac{21}{2}\right)$$

$$= \left(\frac{6}{7} \times \frac{3}{7}\right) - \left(\frac{3}{14} \times \frac{2}{21}\right)$$

$$= \frac{18}{49} - \left(\frac{1}{7} \times \frac{1}{7}\right)$$

$$= \frac{18}{49} - \frac{1}{49}$$

$$= \frac{17}{49}$$

# Decimals to Fractions

## Multiple Choice

**1** **Option C**

$$\frac{0.45}{1} \times \frac{100}{100}$$

$$= \frac{45}{100} = \frac{5 \times 9}{5 \times 20}$$

$$= \frac{9}{20}$$

**2** **Option B**

$$\frac{0.66}{1} \times \frac{100}{100}$$

$$= \frac{66}{100} = \frac{2 \times 33}{2 \times 50}$$

$$= \frac{33}{50}$$

**3** **Option D**

$$\frac{0.7}{1} \times \frac{70}{100} = \frac{7}{10}$$

$$\frac{21}{30} = \frac{70}{100} = \frac{7}{10}$$

$$\frac{3}{4} \neq 0.7$$

**4** **Option B**

$$\frac{1.5}{1} \times \frac{10}{10}$$

$$= \frac{15}{10} = \frac{5 \times 3}{5 \times 2}$$

$$= \frac{3}{2}$$

**5** **Option A**

$$\frac{2.25}{1} \times \frac{100}{100} = \frac{225}{100}$$

$$= \frac{25 \times 9}{25 \times 4}$$

$$= \frac{9}{4}$$

## Time To Practice

**1** **a** 0.3

$$= \frac{0.3}{1} \times \frac{10}{10}$$

$$= \frac{3}{10}$$

**b** 0.7

$$= \frac{0.7}{1} \times \frac{10}{10}$$

$$= \frac{7}{10}$$

**c** 0.8

$$= \frac{0.8}{1} \times \frac{10}{10}$$

$$= \frac{8}{10} = \frac{4}{5}$$

**d** 0.12

$$= \frac{0.12}{1} \times \frac{100}{100}$$

$$= \frac{12}{100} = \frac{4 \times 3}{4 \times 25}$$

$$= \frac{3}{25}$$

**e** 0.13

$$= \frac{0.13}{1} \times \frac{100}{100}$$

$$= \frac{13}{100}$$

**f** 0.22

$$= \frac{0.22}{1} \times \frac{100}{100}$$

$$= \frac{22}{100} = \frac{2 \times 11}{2 \times 50}$$

$$= \frac{11}{50}$$

**g** 0.31

$$= \frac{0.31}{1} \times \frac{100}{100}$$

$$= \frac{31}{100}$$

**h** 0.42

$$= \frac{0.42}{1} \times \frac{100}{100}$$

$$= \frac{42}{100} = \frac{2 \times 21}{2 \times 50}$$

$$= \frac{21}{50}$$

**i** 0.55

$$= \frac{0.55}{1} \times \frac{100}{100}$$

$$= \frac{55}{100} = \frac{5 \times 11}{5 \times 20}$$

$$= \frac{11}{20}$$

**1** **j** 0.62

$$= \frac{0.62}{1} \times \frac{100}{100}$$

$$= \frac{62}{100} = \frac{2 \times 31}{2 \times 50}$$

$$= \frac{31}{50}$$

**k** 0.78

$$= \frac{0.78}{1} \times \frac{100}{100}$$

$$= \frac{78}{100} = \frac{2 \times 39}{2 \times 50}$$

$$= \frac{39}{50}$$

**l** 0.82

$$= \frac{0.82}{1} \times \frac{100}{100}$$

$$= \frac{82}{100} = \frac{2 \times 41}{2 \times 50}$$

$$= \frac{41}{50}$$

**m** 0.88

$$= \frac{0.88}{1} \times \frac{100}{100}$$

$$= \frac{88}{100} = \frac{4 \times 22}{4 \times 25}$$

$$= \frac{22}{25}$$

**n** 0.92

$$= \frac{0.92}{1} \times \frac{100}{100}$$

$$= \frac{92}{100} = \frac{4 \times 23}{4 \times 25}$$

$$= \frac{23}{25}$$

**o** 0.95

$$= \frac{0.95}{1} \times \frac{100}{100}$$

$$= \frac{95}{100} = \frac{5 \times 19}{5 \times 20}$$

$$= \frac{19}{20}$$

**2** **a** 1.3

$$= \frac{1.3}{1} \times \frac{10}{10}$$

$$= \frac{13}{10}$$

**b** 1.4

$$= \frac{1.4}{1} \times \frac{10}{10}$$

$$= \frac{14}{10} = \frac{2 \times 7}{2 \times 5}$$

$$= \frac{7}{5}$$

**c** 1.6

$$= \frac{1.6}{1} \times \frac{10}{10}$$

$$= \frac{16}{10} = \frac{2 \times 8}{2 \times 5}$$

$$= \frac{8}{5}$$

**d** 1.9

$$= \frac{1.9}{1} \times \frac{10}{10}$$

$$= \frac{19}{10}$$

**e** 2.5

$$= \frac{2.5}{1} \times \frac{10}{10}$$

$$= \frac{25}{10} = \frac{5 \times 5}{5 \times 2}$$

$$= \frac{5}{2}$$

**f** 3.9

$$= \frac{3.9}{1} \times \frac{10}{10}$$

$$= \frac{39}{10}$$

**g** 1.46

$$= \frac{1.46}{1} \times \frac{100}{100}$$

$$= \frac{146}{100} = \frac{2 \times 73}{2 \times 50}$$

$$= \frac{73}{50}$$

**h** 1.72

$$= \frac{1.72}{1} \times \frac{100}{100}$$

$$= \frac{172}{100} = \frac{4 \times 43}{4 \times 25}$$

$$= \frac{43}{25}$$

**i** 1.85

$$= \frac{1.85}{1} \times \frac{100}{100}$$

$$= \frac{185}{100} = \frac{5 \times 37}{5 \times 20}$$

$$= \frac{37}{20}$$

## Challenge Yourself

1 a $0.03$

$= \frac{0.03}{1} \times \frac{100}{100}$

$= \frac{3}{100}$

b $0.04$

$= \frac{0.04}{1} \times \frac{100}{100}$

$= \frac{4}{100} = \frac{4 \times 1}{4 \times 25}$

$= \frac{1}{25}$

c $0.005$

$= \frac{0.005}{1} \times \frac{1000}{1000}$

$= \frac{5}{1000} = \frac{5 \times 1}{5 \times 200}$

$= \frac{1}{200}$

d $0.123$

$= \frac{0.123}{1} \times \frac{1000}{1000}$

$= \frac{123}{1000}$

e $0.402$

$= \frac{0.402}{1} \times \frac{1000}{1000}$

$= \frac{402}{1000} = \frac{2 \times 201}{2 \times 500}$

$= \frac{201}{500}$

f $0.888$

$= \frac{0.888}{1} \times \frac{1000}{1000}$

$= \frac{888}{1000}$

$= \frac{8 \times 111}{8 \times 125}$

$= \frac{111}{125}$

Exponents
Powers

# Multiplying and Dividing Powers

## Multiple Choice

**1** **Option D**

$x^3 \times x^4$

$= x^{3+4}$

$= x^7$

**2** **Option B**

$a^5 \div a^2$

$= a^{5-2}$

$= a^3$

**3** **Option A**

$3y^2 \times 2y^5$

$= (3 \times 2)y^{2+5}$

$= 6y^7$

**4** **Option C**

$\frac{p^4 \times p^2}{p^3}$

$= \frac{p^{4+2}}{p^3}$

$= \frac{p^6}{p^3}$

$= p^{6-3}$

$= p^3$

**5** **Option B**

$x^0 \times x^4$

$= 1 \times x^4$

$= x^4$

## Time To Practice

**1**

**a** $2^3 \times 2^2$
$= 2^{3+2}$
$= 2^5 = 32$

**b** $3^2 \times 3$
$= 3^{2+1}$
$= 3^3 = 27$

**c** $5^0 \times 5^3$
$= 1 \times 5^3$
$= 5^3 = 125$

**d** $x^6 \times x^0$
$= x^6 \times 1$
$= x^6$

**e** $y^3 \times y^2$
$= y^{3+2}$
$= y^5$

**f** $z^9 \times z^7$
$= z^{9+7}$
$= z^{16}$

**g** $a^3 \times a^4 \times a^6$
$= a^{3+4+6}$
$= a^{13}$

**h** $b^3 \times b^5 \times b^2$
$= b^{3+5+2}$
$= b^{10}$

**i** $c^2 \times c^8 \times c^9$
$= c^{2+8+9}$
$= c^{19}$

**j** $5 \times (2x^2)^0$
$= 5 \times 1$
$= 5$

**k** $(3x^2y)^0 \times 7$
$= 1 \times 7$
$= 7$

**l** $3(2x^2y^3)^0 - 4$
$= 3 \times 1 - 4$
$= -1$

**m** $6a^4 \times 2a^2$
$= 12a^{4+2}$
$= 12a^6$

**m** $10b^6 \times 2b^3$
$= 20b^{6+3}$
$= 20b^9$

**o** $5c^0 \times 7c^5 \times c^2$
$= 35c^{5+2}$
$= 35c^7$

**2**

**a** $7^6 \div 7^4$
$= 7^{6-4}$
$= 7^2 = 49$

**b** $4^{12} \div 4^9$
$= 4^{12-9}$
$= 4^3 = 64$

**c** $10^{17} \div 10^{14}$
$= 10^{17-14}$
$= 10^3 = 1000$

**d** $m^9 \div m^2$
$= m^{9-2}$
$= m^7$

**e** $n^{20} \div n^5$
$= n^{20-5}$
$= n^{15}$

**f** $p^{15} \div p^3$
$= p^{15-3}$
$= p^{12}$

**2** **g** $\frac{5^5}{5^4} \times 5$

$= 5^{5-4} \times 5$

$= 5 \times 5$

$= 25$

**h** $\frac{3^8}{3^6} \times 3^2$

$= 3^{8-6} \times 3^2$

$= 3^2 \times 3^2$

$= 3^{2+2} = 3^4$

$= 81$

**i** $\frac{2^8}{2^5} \times 2^3$

$= 2^{8-5} \times 2^3$

$= 2^3 \times 2^3$

$= 2^{3+3} = 2^6$

$= 64$

**j** $\frac{10x^5}{2x^2} \times 3x^3$

$= 5x^{5-2} \times 3x^3$

$= 5x^3 \times 3x^3$

$= 15x^{3+3}$

$= 15x^6$

**k** $\frac{14y^8 \times y^0}{7y^4}$

$= \frac{14y^8 \times 1}{7y^4}$

$= 2y^{8-4}$

$= 2y^4$

**l** $\frac{50z^6 \times z^6}{5z^2}$

$= \frac{50z^{6+6}}{5z^2}$

$= 10z^{12-2}$

$= 10z^{10}$

**m** $4a^3 \times 3a^4 \div 2a^2$

$= 12a^{3+4} \div 2a^2$

$= 12a^7 \div 2a^2$

$= 6a^{7-2}$

$= 6a^5$

**n** $12b^9 \times 3b^2 \div 4b^0$

$= 36b^{9+2} \div 4$

$= 36b^{11} \div 4$

$= 9b^{11}$

**o** $16c^7 \times 2c^3 \div 8c^2$

$= 32c^{7+3} \div 8c^2$

$= 32c^{10} \div 8c^2$

$= 4c^{10-2}$

$= 4c^8$

## Challenge Yourself

**1** **a** $\dfrac{12u^{11}v^9 \times 3u^2v^3}{18u^6v^5}$

$= \dfrac{36u^{11+2}v^{9+3}}{18u^6v^5}$

$= \dfrac{36u^{13}v^{12}}{18u^6v^5}$

$= 2u^{13-6}v^{12-5}$

$= 2u^7v^7$

**b** $\dfrac{5e^6f^4 \times 8e^4f^3}{10ef^5}$

$= \dfrac{40e^{6+4}f^{4+3}}{10ef^5}$

$= \dfrac{40e^{10}f^7}{10ef^5}$

$= 4e^{10-1}f^{7-5}$

$= 4e^9f^2$

**2** **a** $\dfrac{-3ab^3}{ab} \div \dfrac{b}{a^4b^2}$

$= \dfrac{-3a^{1-1}b^{3-1}}{1} \div \dfrac{b}{a^4b^2}$

$= \dfrac{-3b^2}{1} \times \dfrac{a^4b^2}{b}$

$= \dfrac{-3a^4b^4}{b}$

$= -3a^4b^{4-1}$

$= -3a^4b^3$

**b** $\dfrac{16a^7b^6}{6a^3b^3} \div \dfrac{8a^6b^3}{27a^3b^2}$

$= \dfrac{16a^{7-3}b^{6-3}}{6} \div \dfrac{8a^{6-3}b^{3-2}}{27}$

$= \dfrac{8a^4b^3}{3} \div \dfrac{8a^3b}{27}$

$= \dfrac{8a^4b^3}{3} \times \dfrac{27}{8a^3b}$

$= \dfrac{a^4b^3}{1} \times \dfrac{9}{a^3b}$

$= \dfrac{9a^4b^3}{a^3b}$

$= 9a^{4-3}b^{3-1}$

$= 9ab^2$

**3** **a** $\dfrac{8x^2y \times 3x^5y^2}{6x^2y^2} \times \dfrac{7x^5y^2}{16x^3y}$

$= \dfrac{24x^{2+5}y^{1+2}}{6x^2y^2} \times \dfrac{7x^5y^2}{16x^3y}$

$= \dfrac{24x^7y^3}{6x^2y^2} \times \dfrac{7x^5y^2}{16x^3y}$

$= \dfrac{4x^{7-2}y^{3-2}}{1} \times \dfrac{7x^5y^2}{16x^3y}$

$= \dfrac{4x^5y}{1} \times \dfrac{7x^5y^2}{16x^3y}$

$= \dfrac{28x^{5+5}y^{2+1}}{16x^3y}$

$= \dfrac{28x^{10}y^3}{16x^3y}$

$= \dfrac{28x^{10-3}y^{3-1}}{16}$

$= \dfrac{7x^7y^2}{4}$

**b** $\dfrac{16x^4y^8}{27x^4y^9} \div \dfrac{x^6y^2}{x^5y^5z^3 \times xyz}$

$= \dfrac{16x^4y^8}{27x^4y^9} \div \dfrac{x^6y^2}{x^{5+1}y^{5+1}z^{3+1}}$

$= \dfrac{16x^4y^8}{27x^4y^9} \div \dfrac{x^6y^2}{x^6y^6z^4}$

$= \dfrac{16x^4y^8}{27x^4y^9} \times \dfrac{x^6y^6z^4}{x^6y^2}$

$= \dfrac{16x^{4+6}y^{8+6}z^4}{27x^{4+6}y^{9+2}}$

$= \dfrac{16x^{10}y^{14}z^4}{27x^{10}y^{11}}$

$= \dfrac{16x^{10-10}y^{14-11}z^4}{27}$

$= \dfrac{16y^3z^4}{27}$

# Power of a Power

## Multiple Choice

**1** **Option C**

$$\left(2ab^3\right)^4 = 2^4a^4b^{3\times4}$$

$$= 16a^4b^{12}$$

**2** **Option A**

$$x^2y^2 \times \left(-3x^2y\right)^2$$

$$= x^2y^2 \times (-3)^2x^{2\times2}y^{1\times2}$$

$$= x^2y^2 \times 9x^4y^2$$

$$= 9x^6y^4$$

**3** **Option A**

$$\left(\frac{3x^2y}{4z}\right)^3 = \frac{3^3x^{2\times3}y^3}{4^3z^3}$$

$$= \frac{27x^6y^3}{64z^3}$$

**4** **Option D**

$$\frac{6a^6b^5}{a^5b^3} \times \frac{a^4}{(2b)^2}$$

$$= 6ab^2 \times \frac{a^4}{4b^2}$$

$$= \frac{6a^5b^2}{4b^2} = \frac{3a^5}{2}$$

**5** **Option B**

$$-\frac{3x^6y^3}{xy} \div \left(\frac{x^2y}{y^3}\right)^2 = -\frac{3x^6y^3}{xy} \div \frac{x^{2\times2}y^2}{y^{3\times2}}$$

$$= -\frac{3x^6y^3}{xy} \div \frac{x^4y^2}{y^6}$$

$$= -\frac{3x^6y^3}{xy} \times \frac{y^6}{x^4y^2}$$

$$= 3x^5y^2 \times \frac{y^4}{x^4}$$

$$= -3xy^6$$

## Time To Practice

**1** **a** $\left(a^2\right)^5$

$= a^{2\times5}$

$= a^{10}$

**b** $\left(b^5\right)^2 \times \left(b^2\right)^3$

$= b^{5\times2} \times b^{2\times3}$

$= b^{10} \times b^6 = b^{16}$

**c** $\left(2a^6\right)^3$

$= 2^3 \times a^{6\times3}$

$= 8a^{18}$

**d** $(-3x^4)^3$

$= (-3)^3 \times x^{4\times3}$

$= -27x^{12}$

**e** $\left(x^2y\right)^3 \times \left(yx^3\right)^2$

$= x^{2\times3}y^3 \times y^2x^{3\times2}$

$= x^6y^3 \times y^2x^6$

$= x^{12}y^5$

**f** $\left(xy^2\right)^8 \div x^4y^5$

$= x^8y^{2\times8} \div x^4y^5$

$= x^8y^{16} \div x^4y^5$

$= x^4y^{11}$

**g** $\left(2x^2y^3\right)^2 \times xy^4$

$= 2^2x^{2\times2}y^{3\times2} \times xy^4$

$= 4x^4y^6 \times xy^4$

$= 4x^5y^{10}$

**h** $\left(2x^3y^2\right)^4 \times \left(x^2y\right)^3$

$= 2^4x^{3\times4}y^{2\times4} \times x^{2\times3}y^3$

$= 16x^{12}y^8 \times x^6y^3$

$= 16x^{18}y^{11}$

**i** $\left(2x^2y^3\right)^4 \div (4x^4)^2$

$= 2^4x^{2\times4}y^{3\times4} \div 4^2x^{4\times2}$

$= 16x^8y^{12} \div 16x^8$

$= y^{12}$

**j** $\left(2m^3n\right)^5 \div \left(8mn^2\right)^2$

$= 2^5m^{3\times5}n^5 \div 8^2m^2n^{2\times2}$

$= 32m^{15}n^5 \div 64m^2n^4$

$= \dfrac{m^{13}n}{2}$

**k** $(2p)^5q^6 \div 9p^4q^3$

$= 2^5p^5q^6 \div 9p^4q^3$

$= \dfrac{32pq^3}{9}$

**l** $r^6s^2 \times 2(3r)^3s^5$

$= r^6s^2 \times 2 \times 3^3r^3s^5$

$= r^6s^2 \times 54r^3s^5$

$= 54r^9s^7$

**2** **a** $\left(\frac{a}{b^2}\right)^2$

$= \frac{a^2}{b^4}$

**b** $\left(\frac{a^2}{b^3}\right)^5$

$= \frac{a^{2\times5}}{b^{3\times5}}$

$= \frac{a^{10}}{b^{15}}$

**c** $\left(\frac{4a^3}{b^2}\right)^2$

$= \frac{4^2a^6}{b^{2\times2}}$

$= \frac{16a^6}{b^4}$

**d** $\left(\frac{x^2}{y^3}\right)^4 \times \left(\frac{x^3}{y}\right)^5$

$= \frac{x^{2\times4}}{y^{3\times4}} \times \frac{x^{3\times5}}{y^5}$

$= \frac{x^8}{y^{12}} \times \frac{x^{15}}{y^5}$

$= \frac{x^{23}}{y^{17}}$

**e** $\left(\frac{2x^3}{y^2}\right)^2 \times \left(\frac{x^3}{2y}\right)^4$

$= \frac{2^2x^{3\times2}}{y^{2\times2}} \times \frac{x^{3\times4}}{2^4y^4}$

$= \frac{4x^6}{y^4} \times \frac{x^{12}}{16y^4}$

$= \frac{x^6}{y^4} \times \frac{x^{12}}{4y^4}$

$= \frac{x^{18}}{4y^8}$

**f** $\left(\frac{3x^2}{2y^3}\right)^3 \div \frac{6x}{y^2}$

$= \frac{3^3x^{2\times3}}{2^3y^{3\times3}} \div \frac{6x}{y^2}$

$= \frac{27x^6}{8y^9} \times \frac{y^2}{6x}$

$= \frac{9x^6}{8y^9} \times \frac{y^2}{2x}$

$= \frac{9x^6y^2}{16xy^9}$

$= \frac{9x^5}{16y^7}$

**2** **g** $\left(\dfrac{6a^4}{b^6}\right)^2 \div \left(\dfrac{3a}{b^2}\right)^4$

$= \dfrac{6^2a^{4\times2}}{b^{6\times2}} \div \dfrac{3^4a^4}{b^{2\times4}}$

$= \dfrac{36a^8}{b^{12}} \div \dfrac{81a^4}{b^8}$

$= \dfrac{36a^8}{b^{12}} \times \dfrac{b^8}{81a^4}$

$= \dfrac{4a^8}{b^{12}} \times \dfrac{b^8}{9a^4}$

$= \dfrac{4a^8b^8}{9a^4b^{12}}$

$= \dfrac{4a^4}{9b^4}$

**h** $\left(\dfrac{9a^4}{2b^2}\right)^2 \div \left(\dfrac{3b^2}{4a^3}\right)^3$

$= \dfrac{9^2a^{4\times2}}{2^2b^{2\times2}} \div \dfrac{3^3b^{2\times3}}{4^3a^{3\times3}}$

$= \dfrac{81a^8}{4b^4} \div \dfrac{27b^6}{64a^9}$

$= \dfrac{81a^8}{4b^4} \times \dfrac{64a^9}{27b^6}$

$= \dfrac{3a^8}{b^4} \times \dfrac{16a^9}{b^6}$

$= \dfrac{48a^{17}}{b^{10}}$

**i** $\left(\dfrac{3a^5}{4b^7}\right)^2 \div 4\left(\dfrac{1}{a^4b^4}\right)^7$

$= \dfrac{3^2a^{5\times2}}{4^2b^{7\times2}} \div 4\left(\dfrac{1}{a^{4\times7}b^{4\times7}}\right)$

$= \dfrac{9a^{10}}{16b^{14}} \div \dfrac{4}{a^{28}b^{28}}$

$= \dfrac{9a^{10}}{16b^{14}} \times \dfrac{a^{28}b^{28}}{4}$

$= \dfrac{9a^{38}b^{28}}{64b^{14}}$

$= \dfrac{9a^{38}b^{14}}{64}$

**j** $\left(\dfrac{3m^4}{n^3}\right)^2 \div \left(\dfrac{n^7}{2m^2}\right)^3$

$= \dfrac{3^2m^{4\times2}}{n^{3\times2}} \div \left(\dfrac{n^{7\times3}}{2^3m^{2\times3}}\right)$

$= \dfrac{9m^8}{n^6} \div \dfrac{n^{21}}{8m^6}$

$= \dfrac{9m^8}{n^6} \times \dfrac{8m^6}{n^{21}}$

$= \dfrac{72m^{14}}{n^{27}}$

**2** **k** $\frac{1}{2}\left(\frac{p^2q}{q^5}\right)^3 \times \left(\frac{p}{q^8}\right)^3$

$= \frac{1}{2}\left(\frac{p^{2\times3}q^3}{q^{5\times3}}\right) \times \frac{p^3}{q^{8\times3}}$

$= \frac{p^6q^3}{2q^{15}} \times \frac{p^3}{q^{24}}$

$= \frac{p^9q^3}{2q^{39}}$

$= \frac{p^9}{2q^{36}}$

**l** $\frac{7}{8}\left(\frac{2r}{3r^3}\right)^3 \div \left(\frac{r^4}{s^2}\right)^3$

$= \frac{7}{8}\left(\frac{2^3r^3}{3^3r^{3\times3}}\right) \div \frac{r^{4\times3}}{s^{2\times3}}$

$= \frac{7}{8}\left(\frac{8r^3}{27r^9}\right) \div \frac{r^{12}}{s^6}$

$= \frac{7}{8}\left(\frac{8r^3}{27r^9}\right) \times \frac{s^6}{r^{12}}$

$= \frac{7}{1}\left(\frac{1}{27r^6}\right) \times \frac{s^6}{r^{12}}$

$= \frac{7}{27r^6} \times \frac{s^6}{r^{12}}$

$= \frac{7s^6}{27r^{18}}$

**3** **a** $\left(\left(x^3\right)^4\right)^2$

$= x^{3\times4\times2}$

$= x^{24}$

**b** $\left(\left(2x^3\right)^2\right)^2$

$= 2^{1\times2\times2}x^{3\times2\times2}$

$= 2^4x^{12}$

$= 16x^{12}$

**c** $\left(\left(x^2y\right)^6\right)^4$

$= x^{2\times6\times4}y^{1\times6\times4}$

$= x^{48}y^{24}$

**d** $\left(\left(\frac{a^3}{b^2}\right)^2\right)^4$

$= \frac{a^{3\times2\times4}}{b^{2\times2\times4}}$

$= \frac{a^{24}}{b^{16}}$

**3** **e** $$\left(\left(\frac{a^3}{b^4c}\right)^2\right)^3$$

$$= \frac{a^{3\times2\times3}}{b^{4\times2\times3}c^{1\times2\times3}}$$

$$= \frac{a^{18}}{b^{24}c^6}$$

**f** $$\left(\left(\frac{a^9}{3b^7}\right)^2\right)^2$$

$$= \frac{a^{9\times2\times2}}{3^{1\times2\times2}b^{7\times2\times2}}$$

$$= \frac{a^{36}}{3^4b^{28}}$$

$$= \frac{a^{36}}{81b^{28}}$$

## Challenge Yourself

**1** **a** $$\left(\frac{4a^3b^2c^4}{2a^2bc}\right)^2 \times \left(\frac{3a^2b}{c^4}\right)^2$$

$$= \left(\frac{2abc^3}{1}\right)^2 \times \left(\frac{3a^2b}{c^4}\right)^2$$

$$= \frac{2^2a^2b^2c^6}{1} \times \frac{3^2a^4b^2}{c^8}$$

$$= \frac{4a^2b^2c^6}{1} \times \frac{9a^4b^2}{c^8}$$

$$= \frac{36a^6b^4c^6}{c^8}$$

$$= \frac{36a^6b^4}{c^2}$$

**b** $$\frac{(4a^2b^2)^2 \times -2(a^4b)^2}{(-3a^3b^2)^2}$$

$$= \frac{4^2a^4b^4 \times -2a^8b^2}{(-3)^2a^6b^4}$$

$$= \frac{16a^4b^4 \times -2a^8b^2}{9a^6b^4}$$

$$= \frac{-32a^{12}b^6}{9a^6b^4}$$

$$= \frac{-32a^6b^2}{9}$$

**2** **a** $\dfrac{(3x^4)^3}{(2y^2z^3)^2} \times \left(\dfrac{2xz}{x^3z}\right)^2$

$= \dfrac{3^3x^{12}}{2^2y^4z^6} \times \dfrac{2^2x^2z^2}{x^6z^2}$

$= \dfrac{27x^{12}}{4y^4z^6} \times \dfrac{4x^2z^2}{x^6z^2}$

$= \dfrac{27x^{12}}{y^4z^6} \times \dfrac{1}{x^4}$

$= \dfrac{27x^8}{y^4z^6}$

**b** $\dfrac{(-5x^3y^2)^3}{(z^2)^2} \times \left(\dfrac{2x^3z}{5y^3}\right)^2$

$= \dfrac{(-5)^3x^9y^6}{z^4} \times \dfrac{2^2x^6z^2}{5^2y^6}$

$= \dfrac{-5x^9}{z^2} \times \dfrac{4x^6}{1}$

$= \dfrac{-20x^{15}}{z^2}$

**3** **a** $\left(\dfrac{-9m^3n}{2mp^3}\right)^2 \div \left(\dfrac{-3mn^3}{p^2}\right)^2$

$= \dfrac{(-9)^2m^6n^2}{2^2m^2p^6} \div \dfrac{(-3)^2m^2n^6}{p^4}$

$= \dfrac{81m^4n^2}{4p^6} \div \dfrac{9m^2n^6}{p^4}$

$= \dfrac{81m^4n^2}{4p^6} \times \dfrac{p^4}{9m^2n^6}$

$= \dfrac{9m^2}{4p^2} \times \dfrac{1}{n^4}$

$= \dfrac{9m^2}{4p^2n^4}$

**b** $\dfrac{6}{7}\left(\dfrac{7r^4st^4}{2s^4t^3}\right)^2 \div \left(\dfrac{-6rs}{r^2}\right)^2$

$= \dfrac{6}{7}\left(\dfrac{7^2r^8s^2t^8}{2^2s^8t^6}\right) \div \dfrac{(-6)^2r^2s^2}{r^4}$

$= \dfrac{6}{7}\left(\dfrac{49r^8t^2}{4s^6}\right) \div \dfrac{36s^2}{r^2}$

$= \dfrac{6}{7}\left(\dfrac{49r^8t^2}{4s^6}\right) \times \dfrac{r^2}{36s^2}$

$= \dfrac{7r^8t^2}{4s^6} \times \dfrac{r^2}{6s^2}$

$= \dfrac{7r^{10}t^2}{24s^8}$

# Negative Powers

## Multiple Choice

**1** **Option B**

$$\frac{1}{8} = \frac{1}{2^3}$$

$$= 2^{-3}$$

**2** **Option C**

$$\frac{1}{x^{-4}} = x^4$$

**3** **Option C**

$$\left(2a^6b^{-4}\right)^{-3} = \frac{1}{(2a^6b^{-4})^3}$$

$$= \frac{1}{2^3a^{18}b^{-12}}$$

$$= \frac{b^{12}}{8a^{18}}$$

**4** **Option B**

$$\left(\frac{3^4}{2^3}\right)^{-2}$$

$$= \left(\frac{2^3}{3^4}\right)^2$$

$$= \frac{2^6}{3^8}$$

**5** **Option A**

$$\left(\frac{x^2y}{y^{-3}}\right)^2 = \left(\frac{x^2yy^3}{1}\right)^2$$

$$= \left(x^2y^4\right)^2$$

$$= x^4y^8$$

## Time To Practice

**1** **a** $4^{-1}$

$= \frac{1}{4}$

**b** $3^{-3}$

$= \frac{1}{3^3}$

$= \frac{1}{27}$

**c** $5^{-3}$

$= \frac{1}{5^3}$

$= \frac{1}{125}$

**d** $2^{-3} \times 2^2$

$= 2^{-3+2} = 2^{-1}$

$= \frac{1}{2}$

**e** $7^{-4} \times 7^2$

$= 7^{-4+2} = 7^{-2}$

$= \frac{1}{7^2}$

$= \frac{1}{49}$

**f** $\frac{3}{4^{-2}}$

$= 3 \times \frac{4^2}{1}$

$= 3 \times 16$

$= 48$

**g** $\frac{1}{a^{-3}}$

$= a^3$

**h** $b^{-4} \times b$

$= b^{-4+1} = b^{-3}$

$= \frac{1}{b^3}$

**i** $c^{-5} \div c^{-3}$

$= c^{-5-(-3)}$

$= c^{-5+3} = c^{-2}$

$= \frac{1}{c^2}$

**j** $\frac{1}{y^{-4} \times y}$

$= \frac{1}{y^{-4+1}} = \frac{1}{y^{-3}}$

$= y^3$

**k** $x^{-3} \times y \times x^2$

$= x^{-3+2}y$

$= x^{-1}y$

$= \frac{1}{x} \times y$

$= \frac{y}{x}$

**l** $x^{-4} \times y^{-3} \times z^{-2}$

$= \frac{1}{x^4} \times \frac{1}{y^3} \times \frac{1}{z^2}$

$= \frac{1}{x^4y^3z^2}$

**2** **a** $\left(\frac{5}{4}\right)^{-3}$

$= \left(\frac{4}{5}\right)^{3}$

$= \frac{4^3}{5^3}$

$= \frac{64}{125}$

**b** $\left(\frac{2}{3}\right)^{-2}$

$= \left(\frac{3}{2}\right)^{2}$

$= \frac{3^2}{2^2}$

$= \frac{9}{4}$

**c** $\left(\frac{1}{2^{-3}}\right)^{-2}$

$= (2^{-3})^2$

$= 2^{-6}$

$= \frac{1}{2^6}$

$= \frac{1}{64}$

**d** $\left(\frac{x^2}{y^{-3}}\right)^{4}$

$= \left(\frac{x^2y^3}{1}\right)^{4}$

$= x^8y^{12}$

**e** $\left(\frac{2x^{-3}}{y^{-2}}\right)^{-1}$

$= \left(\frac{2y^2}{x^3}\right)^{-1}$

$= \left(\frac{x^3}{2y^2}\right)^{1}$

$= \frac{x^3}{2y^2}$

**f** $\left(\frac{2}{x}\right)^{-3} \times x^{-2}$

$= \left(\frac{x}{2}\right)^{3} \times \frac{1}{x^2}$

$= \frac{x^3}{8} \times \frac{1}{x^2}$

$= \frac{x}{8}$

**g** $\frac{p^{-3}q^2}{r^{-4}p^2}$

$= \frac{q^2r^4}{p^3p^2}$

$= \frac{q^2r^4}{p^5}$

**h** $\frac{(p^{-1})^3 \times (q^3)^{-3}}{p^{-6} \times (q^{-5})^3}$

$= \frac{p^{-3} \times q^{-9}}{p^{-6} \times q^{-15}}$

$= \frac{p^6 \times q^{15}}{p^3 \times q^9}$

$= p^3q^6$

**i** $\frac{(p^2)^{-2} \times (q^{-3})^{-2}}{p^4 \times (q^{-1})^2}$

$= \frac{p^{-4} \times q^6}{p^4 \times q^{-2}}$

$= \frac{q^6 \times q^2}{p^4 \times p^4}$

$= \frac{q^8}{p^8}$

**2** **j** $\frac{x^{-3}y^{-4}}{z^{-2}} \times \frac{x^{-2}y^{3}}{z^{-2}}$

$= \frac{z^2}{x^3y^4} \times \frac{y^3z^2}{x^2}$

$= \frac{z^4}{x^5y}$

**k** $\left(\frac{x^{-2}y}{z^2}\right)^3 \times \left(\frac{z}{x^{-1}}\right)^{-3}$

$= \left(\frac{y}{x^2z^2}\right)^3 \times \left(\frac{xz}{1}\right)^{-3}$

$= \left(\frac{y}{x^2z^2}\right)^3 \times \left(\frac{1}{xz}\right)^3$

$= \frac{y^3}{x^6z^6} \times \frac{1}{x^3z^3}$

$= \frac{y^3}{x^9z^9}$

**l** $\left(\frac{rs}{r^{-5}}\right)^{-3} \div \left(\frac{2r^{-8}}{s^2}\right)^{-2}$

$= \left(\frac{rsr^5}{1}\right)^{-3} \div \left(\frac{2}{r^8s^2}\right)^{-2}$

$= \left(\frac{1}{r^6s}\right)^3 \div \left(\frac{r^8s^2}{2}\right)^2$

$= \frac{1}{r^{18}s^3} \div \frac{r^{16}s^4}{4}$

$= \frac{1}{r^{18}s^3} \times \frac{4}{r^{16}s^4}$

$= \frac{4}{r^{34}s^7}$

**3** **a** $\left(\frac{x^{-3}}{2^{-2}y^{-4}}\right)^{-2}$

$= \left(\frac{2^2y^4}{x^3}\right)^{-2}$

$= \left(\frac{x^3}{4y^4}\right)^{2}$

$= \frac{x^6}{4^2y^8}$

$= \frac{x^6}{16y^8}$

**b** $\left(\frac{m^{-2}n^{-3}}{3p^2}\right)^{-2} \div p^{-6}$

$= \left(\frac{1}{3p^2m^2n^3}\right)^{-2} \div \frac{1}{p^6}$

$= \left(\frac{3p^2m^2n^3}{1}\right)^{2} \div \frac{1}{p^6}$

$= \frac{9p^4m^4n^6}{1} \div \frac{1}{p^6}$

$= \frac{9p^4m^4n^6}{1} \times \frac{p^6}{1}$

$= 9p^{10}m^4n^6$

**c** $\left(\frac{2^{-2}x^{-1}}{y^3}\right)^{-2} \div y^{-9}$

$= \left(\frac{1}{2^2xy^3}\right)^{-2} \div \frac{1}{y^9}$

$= \left(\frac{2^2xy^3}{1}\right)^{2} \div \frac{1}{y^9}$

$= \frac{2^4x^2y^6}{1} \div \frac{1}{y^9}$

$= \frac{16x^2y^6}{1} \times \frac{y^9}{1}$

$= 16x^2y^{15}$

## Challenge Yourself

**1** **a** $\dfrac{2^{-4}x^{-3}y^2}{z^{-2}} \times \dfrac{x^6y^4}{z^{-4}}$

$= \dfrac{y^2z^2}{2^4x^3} \times \dfrac{x^6y^4z^4}{1}$

$= \dfrac{x^6y^6z^6}{16x^3}$

$= \dfrac{x^3y^6z^6}{16}$

**b** $\dfrac{2^{-3}x^2y^{-4}}{z^{-3}} \times \left(\dfrac{x^{-1}z}{y^{-2}}\right)^{-2}$

$= \dfrac{x^2z^3}{2^3y^4} \times \left(\dfrac{y^2z}{x}\right)^{-2}$

$= \dfrac{x^2z^3}{8y^4} \times \left(\dfrac{x}{y^2z}\right)^{2}$

$= \dfrac{x^2z^3}{8y^4} \times \dfrac{x^2}{y^4z^2}$

$= \dfrac{x^4z}{8y^8}$

**2** **a** $\dfrac{(m^{-4})^3 \times (n^{-3})^2}{(2m^2n^3)^2} \times \left(\dfrac{mn}{3m^{-3}}\right)^{-2}$

$= \dfrac{m^{-12} \times n^{-6}}{2^2m^4n^6} \times \left(\dfrac{mnm^3}{3}\right)^{-2}$

$= \dfrac{1}{2^2m^4n^6m^{12}n^6} \times \left(\dfrac{3}{m^4n}\right)^{2}$

$= \dfrac{1}{4m^{16}n^{12}} \times \dfrac{9}{m^8n^2}$

$= \dfrac{9}{4m^{24}n^{14}}$

**b** $\dfrac{4}{7}\left(\dfrac{m^3n^{-4}}{p^{-2}}\right)^{-2} \times \left(\dfrac{p^2m^{-1}}{n^3}\right)^{-3}$

$= \dfrac{4}{7}\left(\dfrac{m^3p^2}{n^4}\right)^{-2} \times \left(\dfrac{p^2}{n^3m}\right)^{-3}$

$= \dfrac{4}{7}\left(\dfrac{n^4}{m^3p^2}\right)^{2} \times \left(\dfrac{n^3m}{p^2}\right)^{3}$

$= \dfrac{4}{7}\left(\dfrac{n^8}{m^6p^4}\right) \times \dfrac{n^9m^3}{p^6}$

$= \dfrac{4n^8}{7m^3p^4} \times \dfrac{n^9}{p^6}$

$= \dfrac{4n^{17}}{7m^3p^{10}}$

**3** **a** $\frac{2}{5}\left(\frac{3x^{-3}y^{-1}}{x^{-2}y^{-4}z}\right)^{-2} \div \left(\frac{x^2y^{-1}}{4y^5z^3}\right)^{-1}$

$= \frac{2}{5}\left(\frac{3x^2y^4}{x^3yz}\right)^{-2} \div \left(\frac{x^2}{4y^5zy}\right)^{-1}$

$= \frac{2}{5}\left(\frac{3x^2y^4}{x^3yz}\right)^{-2} \div \left(\frac{x^2}{4y^6z}\right)^{-1}$

$= \frac{2}{5}\left(\frac{3y^3}{xz}\right)^{-2} \div \left(\frac{x^2}{4y^6z}\right)^{-1}$

$= \frac{2}{5}\left(\frac{xz}{3y^3}\right)^{2} \div \left(\frac{4y^6z}{x^2}\right)^{1}$

$= \frac{2}{5}\left(\frac{x^2z^2}{9y^6}\right) \div \frac{4y^6z}{x^2}$

$= \frac{2x^2z^2}{45y^6} \times \frac{x^2}{4y^6z}$

$= \frac{x^2}{45y^6} \times \frac{x^2}{2y^6z}$

$= \frac{x^4}{90y^{12}z}$

**b** $\left(\frac{x^{-2}y^{-1}}{x^{-4}y^{-1}z^{-5}}\right)^{-2} \div \left(\frac{x^{-2}y^{-5}z^{-9}}{3^{-1}y^{-8}z^{-1}}\right)^{-1}$

$= \left(\frac{x^4yz^5}{x^2y}\right)^{-2} \div \left(\frac{3y^8z}{x^2y^5z^9}\right)^{-1}$

$= \left(\frac{x^4z^5}{x^2}\right)^{-2} \div \left(\frac{3y^3}{x^2z^8}\right)^{-1}$

$= \left(\frac{x^2}{x^4z^5}\right)^{2} \div \left(\frac{x^2z^8}{3y^3}\right)^{1}$

$= \frac{x^4}{x^8z^{10}} \times \frac{3y^3}{x^2z^8}$

$= \frac{1}{x^4z^{10}} \times \frac{3y^3}{x^2z^8}$

$= \frac{3y^3}{x^6z^{18}}$

# Fractional Powers

## Multiple Choice

1 **Option A**

$16^{\frac{1}{4}}$

$= (2^4)^{\frac{1}{4}}$

$= 2^1$

$= 2$

2 **Option C**

$32^{\frac{3}{5}}$

$= \left(32^{\frac{1}{5}}\right)^3$

$= (2)^3$

$= 8$

3 **Option B**

$\left(\frac{64}{x^3}\right)^{\frac{1}{3}}$

$= \frac{64^{\frac{1}{3}}}{x^1}$

$= \frac{4}{x}$

4 **Option D**

$(81y^{16})^{\frac{1}{4}}$

$= 81^{\frac{1}{4}}y^4$

$= 3y^4$

5 **Option B**

$(a^9b^6)^{\frac{2}{3}}$

$= a^{\frac{18}{3}}b^{\frac{12}{3}}$

$= a^6b^4$

## Time To Practice

**1** **a** $27^{\frac{1}{3}}$

$= 3$

**b** $121^{\frac{1}{2}}$

$= 11$

**c** $49^{\frac{1}{2}}$

$= 7$

**d** $8^{\frac{5}{3}}$

$= \left(8^{\frac{1}{3}}\right)^5 = (2)^5 = 32$

**e** $25^{\frac{3}{2}}$

$= \left(25^{\frac{1}{2}}\right)^3$

$= (5)^3 = 125$

**f** $49^{\frac{3}{2}}$

$= \left(49^{\frac{1}{2}}\right)^3$

$= (7)^3 = 343$

**g** $16^{\frac{3}{4}}$

$= \left(16^{\frac{1}{4}}\right)^3$

$= (2)^3 = 8$

**h** $4^{-\frac{1}{2}}$

$= \frac{1}{4^{\frac{1}{2}}}$

$= \frac{1}{2}$

**i** $27^{-\frac{2}{3}}$

$= \frac{1}{27^{\frac{2}{3}}} = \frac{1}{\left(27^{\frac{1}{3}}\right)^2}$

$= \frac{1}{(3)^2} = \frac{1}{9}$

**j** $8^{-\frac{1}{3}}$

$= \frac{1}{8^{\frac{1}{3}}} = \frac{1}{2}$

**k** $64^{-\frac{2}{3}}$

$= \frac{1}{64^{\frac{2}{3}}} = \frac{1}{\left(64^{\frac{1}{3}}\right)^2}$

$= \frac{1}{(4)^2} = \frac{1}{16}$

**l** $(125^{-1})^{\frac{1}{3}}$

$= \left(\frac{1}{125}\right)^{\frac{1}{3}}$

$= \frac{1}{125^{\frac{1}{3}}} = \frac{1}{5}$

**2** **a** $\left(\frac{9}{16}\right)^{-\frac{1}{2}}$

$= \left(\frac{16}{9}\right)^{\frac{1}{2}}$

$= \frac{16^{\frac{1}{2}}}{9^{\frac{1}{2}}} = \frac{4}{3}$

**b** $\left(\frac{16}{81}\right)^{-\frac{3}{4}}$

$= \left(\frac{81}{16}\right)^{\frac{3}{4}} = \frac{81^{\frac{3}{4}}}{16^{\frac{3}{4}}}$

$= \frac{\left(81^{\frac{1}{4}}\right)^3}{\left(16^{\frac{1}{4}}\right)^3}$

$= \frac{3^3}{2^3} = \frac{27}{8}$

**c** $\left(\frac{27}{64}\right)^{-\frac{2}{3}}$

$= \left(\frac{64}{27}\right)^{\frac{2}{3}} = \frac{64^{\frac{2}{3}}}{27^{\frac{2}{3}}}$

$= \frac{\left(64^{\frac{1}{3}}\right)^2}{\left(27^{\frac{1}{3}}\right)^2} = \frac{4^2}{3^2} = \frac{16}{9}$

**3** **a** $\sqrt{x^3}$

$= x^{\frac{2}{3}}$

**b** $\sqrt[3]{x^5}$

$= x^{\frac{5}{3}}$

**c** $\sqrt[4]{16x}$

$= (16x)^{\frac{1}{4}}$

**d** $\sqrt[3]{y^2}$

$= \left(y^2\right)^{\frac{1}{3}} = y^{\frac{2}{3}}$

**e** $\sqrt[6]{64y^7}$

$= (64y^7)^{\frac{1}{6}}$

**f** $\sqrt[3]{8y^4}$

$= (8y^4)^{\frac{1}{3}}$

4

a $(x^6)^{\frac{1}{3}}$

$= x^{\frac{6}{3}} = x^2$

b $(25x^2)^{\frac{1}{2}}$

$= 25^{\frac{1}{2}}x = 5x$

c $(16x^8)^{\frac{1}{4}}$

$16^{\frac{1}{4}}x^2 = 2x^2$

d $(a^2b^4)^{\frac{1}{2}}$

$= ab^2$

e $(8a^3b^3)^{\frac{1}{3}}$

$= 8^{\frac{1}{3}}ab$

$= 2ab$

f $3a^{\frac{1}{2}} \times 2a^{-\frac{3}{4}}$

$= 6a^{\frac{1}{2}+(-\frac{3}{4})}$

$= 6a^{\frac{1}{2}+(-\frac{3}{4})}$

$= 6a^{-\frac{1}{4}}$

$= \frac{6}{a^{\frac{1}{4}}}$

g $(a^8)^{-\frac{1}{2}}$

$= a^{-4}$

$= \frac{1}{a^4}$

h $(8x^9)^{-\frac{1}{3}}$

$= 8^{-\frac{1}{3}}a^{-3}$

$= \frac{1}{8^{\frac{1}{3}}a^3}$

$= \frac{1}{2a^3}$

i $(x^4y^8)^{-\frac{3}{4}}$

$= x^{-3}y^{-6}$

$= \frac{1}{x^3y^6}$

j $(x^2)^{\frac{1}{2}} \times x^{-\frac{5}{3}}$

$= x \times x^{-\frac{5}{3}}$

$= x^{1+(-\frac{5}{3})}$

$= x^{-\frac{2}{3}}$

$= \frac{1}{x^{\frac{2}{3}}}$

k $\left(\frac{x^{12}y^{-8}}{z^4}\right)^{\frac{3}{4}}$

$= \left(\frac{x^{12}}{y^8z^4}\right)^{\frac{3}{4}}$

$= \frac{x^9}{y^6z^3}$

l $\left(\frac{x^6}{y^{-9}}\right)^{\frac{2}{3}}$

$= (x^6y^9)^{\frac{2}{3}}$

$= x^4y^6$

## Challenge Yourself

**1** **a** $\left(a^2b^4\right)^{-\frac{1}{2}} \times \left(ba^2\right)^{\frac{3}{2}}$

$= a^{-1}b^{-2} \times b^{\frac{3}{2}}a^3$

$= a^{-1+3}b^{-2+\frac{3}{2}}$

$= a^2b^{-\frac{1}{2}}$

$= \dfrac{a^2}{b^{\frac{1}{2}}}$

**b** $\left(a^3b^{-2}\right)^{-\frac{2}{3}} \times \left(a^{-1}b^2\right)^{\frac{1}{3}}$

$= a^{-2}b^{\frac{4}{3}} \times a^{-\frac{1}{3}}b^{\frac{2}{3}}$

$= a^{-2-\frac{1}{3}}b^{\frac{4}{3}+\frac{2}{3}}$

$= a^{-\frac{7}{3}}b^2$

$= \dfrac{b^2}{a^{\frac{7}{3}}}$

**2** **a** $\left(2^{-1}x^{\frac{3}{2}}y^{\frac{5}{4}}\right)^2 \times \left(8x^{-\frac{7}{4}}y^{\frac{1}{2}}\right)^{\frac{2}{3}}$

$= 2^{-2}x^3y^{\frac{5}{2}} \times 8^{\frac{2}{3}}x^{-\frac{7}{6}}y^{\frac{1}{3}}$

$= \dfrac{1}{2^2} \times \left(8^{\frac{1}{3}}\right)^2 \times x^{3-\frac{7}{6}}y^{\frac{5}{2}+\frac{1}{3}}$

$= \dfrac{1}{4} \times (2)^2 \times x^{\frac{11}{6}}y^{\frac{17}{6}}$

$= x^{\frac{11}{6}}y^{\frac{17}{6}}$

**b** $\left(x^{\frac{2}{5}}y^{-\frac{1}{4}}\right)^2 \div \left(x^{-\frac{3}{2}}y^{\frac{1}{3}}\right)^{\frac{2}{3}}$

$= x^{\frac{4}{5}}y^{-\frac{1}{2}} \div x^{-1}y^{\frac{2}{9}}$

$= x^{\frac{4}{5}-(-1)}y^{-\frac{1}{2}-\frac{2}{9}}$

$= x^{\frac{9}{5}}y^{-\frac{13}{18}}$

$= \dfrac{x^{\frac{9}{5}}}{y^{\frac{13}{18}}}$

**3** **a** $\left(\dfrac{a^{\frac{1}{2}}b^{\frac{1}{3}}}{2c^{-\frac{1}{2}}}\right)^2 \times \left(a^{-1}c^{\frac{1}{3}}\right)^{-1}$

$= \dfrac{ab^{\frac{2}{3}}}{4c^{-1}} \times ac^{-\frac{1}{3}}$

$= \dfrac{ab^{\frac{2}{3}}c}{4} \times ac^{-\frac{1}{3}}$

$= \dfrac{a^2b^{\frac{2}{3}}c^{1-\frac{1}{3}}}{4}$

$= \dfrac{a^2b^{\frac{2}{3}}c^{\frac{2}{3}}}{4}$

**b** $\left(\dfrac{2a^2b^{-1}}{c^{\frac{1}{2}}}\right)^{-2} \div \left(\dfrac{4a^{-1}b^2}{c}\right)^{\frac{1}{2}}$

$= \left(\dfrac{2a^2}{bc^{\frac{1}{2}}}\right)^{-2} \div \left(\dfrac{4b^2}{ac}\right)^{\frac{1}{2}}$

$= \left(\dfrac{bc^{\frac{1}{2}}}{2a^2}\right)^2 \div \dfrac{2b}{a^{\frac{1}{2}}c^{\frac{1}{2}}}$

$= \dfrac{b^2c}{4a^4} \times \dfrac{a^{\frac{1}{2}}c^{\frac{1}{2}}}{2b}$

$= \dfrac{bc^{\frac{3}{2}}}{8a^{\frac{7}{2}}}$

Surds

# Identifying Surds

## Multiple Choice

**1** **Option C**

No number multiplied by itself equals 50, $\therefore \sqrt{50}$ is a surd

**2** **Option B**

$\sqrt[3]{64} = 4$

$\therefore$ Not a surd. Option B is NOT true.

**3** **Option D**

$\sqrt[3]{27} = 3$

$\sqrt[3]{1000} = 10$

$\sqrt[3]{125} = 5$

$\therefore \sqrt[3]{20}$ is a surd.

**4** **Option A**

$$\sqrt{0.001} = \sqrt{\frac{1}{1000}}$$

$$= \sqrt{\frac{1}{100 \times 10}}$$

$$= \frac{1}{10}\sqrt{\frac{1}{10}}$$

$\therefore \sqrt{0.001}$ is a surd.

**5** **Option D**

$$\sqrt{\frac{5}{25}} = \sqrt{\frac{1}{5}}$$

No number multiplied by itself equals 5, $\therefore \sqrt{\frac{5}{25}}$ is a surd.

## Time To Practice

**1** **a** $\sqrt{10}$

No number multiplied by itself equals 10.

$\therefore$ Yes, it's a surd.

**b** $\sqrt{144} = \sqrt{12 \times 12} = 12$

$\therefore$ NOT a surd.

**c** $\sqrt{55}$

No number multiplied by itself equals 55.

$\therefore$ Yes, it's a surd.

**d** $\sqrt{18}$

No number multiplied by itself equals 18.

$\therefore$ Yes, it's a surd.

**e** $\sqrt{8}$

No number multiplied by itself equals 8.

$\therefore$ Yes, it's a surd.

**f** $\sqrt{1} = \sqrt{1 \times 1} = 1$

$\therefore$ NOT a surd.

**g** $\sqrt[3]{21}$

No number multiplied by itself three times equals 21.

$\therefore$ Yes, it's a surd.

**h** $\sqrt[3]{216} = \sqrt[3]{6 \times 6 \times 6} = 6$

$\therefore$ NOT a surd.

**i** $\sqrt[3]{32}$

No number multiplied by itself three times equals 32.

$\therefore$ Yes, it's a surd.

**j** $\sqrt[4]{16} = \sqrt[4]{2 \times 2 \times 2 \times 2} = 2$

$\therefore$ NOT a surd.

**1** **k** $\sqrt[4]{40}$

No number multiplied by itself four times equals 40.

∴ Yes, it's a surd.

**l** $\sqrt[5]{32} = \sqrt[5]{2 \times 2 \times 2 \times 2 \times 2} = 2$

∴ NOT a surd.

**m** $\sqrt{\frac{1}{4}} = \sqrt{\frac{1}{2} \times \frac{1}{2}} = \frac{1}{2}$

∴ NOT a surd.

**n** $\sqrt{\frac{4}{49}} = \sqrt{\frac{2}{7} \times \frac{2}{7}} = \frac{2}{7}$

∴ NOT a surd.

**o** $\sqrt{\frac{16}{32}}$

No number multiplied by itself equals 32.

∴ Yes, it's a surd.

**p** $\sqrt{0.36} = \sqrt{\frac{36}{100}} = \sqrt{\frac{6}{10} \times \frac{6}{10}}$

$= \frac{6}{10} = \frac{3}{5}$

∴ NOT a surd.

**q** $\sqrt{0.5} = \sqrt{\frac{50}{100}}$

No number multiplied by itself two times equals 50.

∴ Yes, it's a surd.

**r** $\sqrt{0.65} = \sqrt{\frac{65}{100}} = \sqrt{\frac{17}{20}}$

No number multiplied by itself two times equals 17 or 20.

∴ Yes, it's a surd.

## Challenge Yourself

**1** **a** $\sqrt[3]{\frac{1}{8}}$

$= \sqrt[3]{\frac{1}{2} \times \frac{1}{2} \times \frac{1}{2}} = \frac{1}{2}$

∴ NOT a surd.

**b** $\sqrt[4]{\frac{1}{32}}$

No number multiplied by itself four times equals 32.

∴ Yes, it's a surd.

**c** $\sqrt[3]{\frac{64}{125}}$

$= \sqrt[3]{\frac{4}{5} \times \frac{4}{5} \times \frac{4}{5}} = \frac{4}{5}$

∴ NOT a surd.

**d** $\sqrt[3]{0.125}$

$= \sqrt[3]{\frac{125}{1000}} = \sqrt[3]{\frac{5}{10} \times \frac{5}{10} \times \frac{5}{10}} = \frac{5}{10}$

∴ NOT a surd.

**e** $\sqrt[3]{0.1} == \sqrt[3]{\frac{1}{10}}$

No number multiplied by itself three times equals 10.

∴ Yes, it's a surd.

**f** $\sqrt[3]{0.008}$

$= \sqrt[3]{\frac{8}{1000}} = \sqrt[3]{\frac{2}{10} \times \frac{2}{10} \times \frac{2}{10}} = \frac{2}{10}$

∴ NOT a surd.

**2**

$\sqrt{\frac{1}{8}}$ → No number multiplied by itself equals 8.

$\sqrt{21}$ → No number multiplied by itself equals 21.

# Simplifying Surds

## Multiple Choice

**1** **Option A**

$\sqrt{50}$

$= \sqrt{25 \times 2}$

$= \sqrt{25} \times \sqrt{2}$

$= 5\sqrt{2}$

**2** **Option C**

$\sqrt{200}$

$= \sqrt{100 \times 2}$

$= \sqrt{100} \times \sqrt{2}$

$= 10\sqrt{2}$

**3** **Option D**

$3\sqrt{27}$

$= 3\sqrt{9 \times 3}$

$= 3\sqrt{9} \times \sqrt{3}$

$= 3 \times 3 \times \sqrt{3}$

$= 9\sqrt{3}$

**4** **Option B**

$\frac{3}{4}\sqrt{80}$

$= \frac{3}{4}\sqrt{16 \times 5}$

$= \frac{3}{4}\sqrt{16} \times \sqrt{5}$

$= \frac{3}{4} \times 4 \times \sqrt{5}$

$= 3\sqrt{5}$

**5** **Option A**

$6\sqrt{3}$

$= \sqrt{36 \times 3}$

$= \sqrt{108}$

$\therefore 108$

## Time To Practice

1 a $\sqrt{12}$

$= \sqrt{4 \times 3} = \sqrt{4} \times \sqrt{3}$

$= 2\sqrt{3}$

b $\sqrt{18}$

$= \sqrt{9 \times 2} = \sqrt{9} \times \sqrt{2}$

$= 3\sqrt{2}$

c $\sqrt{20}$

$= \sqrt{4 \times 5} = \sqrt{4} \times \sqrt{5}$

$= 2\sqrt{5}$

d $\sqrt{24}$

$= \sqrt{4 \times 6} = \sqrt{4} \times \sqrt{6}$

$= 2\sqrt{6}$

e $\sqrt{27}$

$= \sqrt{9 \times 3} = \sqrt{9} \times \sqrt{3}$

$= 3\sqrt{3}$

f $\sqrt{72}$

$= \sqrt{36 \times 2} = \sqrt{36} \times \sqrt{2}$

$= 6\sqrt{2}$

g $\sqrt{98}$

$= \sqrt{49 \times 2} = \sqrt{49} \times \sqrt{2}$

$= 7\sqrt{2}$

h $\sqrt{125}$

$= \sqrt{25 \times 5} = \sqrt{25} \times \sqrt{5}$

$= 5\sqrt{5}$

i $\sqrt{360}$

$= \sqrt{36 \times 10} = \sqrt{36} \times \sqrt{10}$

$= 6\sqrt{10}$

**2** **a** $2\sqrt{18}$

$= 2\sqrt{9 \times 2} = 2\sqrt{9} \times \sqrt{2}$

$= 2 \times 3 \times \sqrt{2}$

$= 6\sqrt{2}$

**b** $3\sqrt{20}$

$= 3\sqrt{4 \times 5} = 3\sqrt{4} \times \sqrt{5}$

$= 3 \times 2 \times \sqrt{5}$

$= 6\sqrt{5}$

**c** $2\sqrt{63}$

$= 2\sqrt{9 \times 7} = 2\sqrt{9} \times \sqrt{7}$

$= 2 \times 3 \times \sqrt{7}$

$= 6\sqrt{7}$

**d** $4\sqrt{48}$

$= 4\sqrt{16 \times 3} = 4\sqrt{16} \times \sqrt{3}$

$= 4 \times 4 \times \sqrt{3}$

$= 16\sqrt{3}$

**e** $5\sqrt{98}$

$= 5\sqrt{49 \times 2} = 5\sqrt{49} \times \sqrt{2}$

$= 5 \times 7 \times \sqrt{2}$

$= 35\sqrt{2}$

**f** $2\sqrt{800}$

$= 2\sqrt{400 \times 2} = 2\sqrt{400} \times \sqrt{2}$

$= 2 \times 20 \times \sqrt{2}$

$= 40\sqrt{2}$

**g** $4\sqrt{150}$

$= 4\sqrt{25 \times 6} = 4\sqrt{25} \times \sqrt{6}$

$= 4 \times 5 \times \sqrt{6}$

$= 20\sqrt{6}$

**h** $3\sqrt{147}$

$= 3\sqrt{49 \times 3} = 3\sqrt{49} \times \sqrt{3}$

$= 3 \times 7 \times \sqrt{3}$

$= 21\sqrt{3}$

**i** $6\sqrt{75}$

$= 6\sqrt{25 \times 3} = 6\sqrt{25} \times \sqrt{3}$

$= 6 \times 5 \times \sqrt{3}$

$= 30\sqrt{3}$

**3** **a** $\frac{1}{2}\sqrt{20}$

$= \frac{1}{2}\sqrt{4 \times 5} = \frac{1}{2}\sqrt{4} \times \sqrt{5}$

$= \frac{1}{2} \times 2 \times \sqrt{5}$

$= \sqrt{5}$

**b** $\frac{1}{3}\sqrt{28}$

$= \frac{1}{3}\sqrt{4 \times 7} = \frac{1}{3}\sqrt{4} \times \sqrt{7}$

$= \frac{1}{3} \times 2 \times \sqrt{7}$

$= \frac{2\sqrt{7}}{3}$

**c** $\frac{1}{4}\sqrt{54}$

$= \frac{1}{4}\sqrt{9 \times 6} = \frac{1}{4}\sqrt{9} \times \sqrt{6}$

$= \frac{1}{4} \times 3 \times \sqrt{6}$

$= \frac{3\sqrt{6}}{4}$

**d** $\frac{1}{6}\sqrt{24}$

$= \frac{1}{6}\sqrt{4 \times 6} = \frac{1}{6}\sqrt{4} \times \sqrt{6}$

$= \frac{1}{6} \times 2 \times \sqrt{6}$

$= \frac{2\sqrt{6}}{6} = \frac{\sqrt{6}}{3}$

**e** $\frac{1}{5}\sqrt{80}$

$= \frac{1}{5}\sqrt{16 \times 5} = \frac{1}{5}\sqrt{16} \times \sqrt{5}$

$= \frac{1}{5} \times 4 \times \sqrt{5}$

$= \frac{4\sqrt{5}}{5}$

**f** $\frac{1}{9}\sqrt{99}$

$= \frac{1}{9}\sqrt{9 \times 11} = \frac{1}{9}\sqrt{9} \times \sqrt{11}$

$= \frac{1}{9} \times 3 \times \sqrt{11}$

$= \frac{3\sqrt{11}}{9} = \frac{\sqrt{11}}{3}$

**3** **g** $\frac{1}{8}\sqrt{84}$

$= \frac{1}{8}\sqrt{4 \times 21} = \frac{1}{8}\sqrt{4} \times \sqrt{21}$

$= \frac{1}{8} \times 2 \times \sqrt{21}$

$= \frac{2\sqrt{21}}{8} = \frac{\sqrt{21}}{4}$

**h** $\frac{1}{7}\sqrt{56}$

$= \frac{1}{7}\sqrt{4 \times 14} = \frac{1}{7}\sqrt{4} \times \sqrt{14}$

$= \frac{1}{7} \times 2 \times \sqrt{14}$

$= \frac{2\sqrt{14}}{7}$

**i** $\frac{1}{3}\sqrt{68}$

$= \frac{1}{3}\sqrt{4 \times 17} = \frac{1}{3}\sqrt{4} \times \sqrt{17}$

$= \frac{1}{3} \times 2 \times \sqrt{17}$

$= \frac{2\sqrt{17}}{3}$

**4** **a** $5\sqrt{2}$

$= \sqrt{5^2 \times 2} = \sqrt{25 \times 2}$

$= \sqrt{50}$

**b** $3\sqrt{3}$

$= \sqrt{3^2 \times 3} = \sqrt{9 \times 3}$

$= \sqrt{27}$

**c** $8\sqrt{2}$

$= \sqrt{8^2 \times 2} = \sqrt{64 \times 2}$

$= \sqrt{128}$

**d** $10\sqrt{7}$

$= \sqrt{10^2 \times 7} = \sqrt{100 \times 7}$

$= \sqrt{700}$

**4** **e** $9\sqrt{10}$

$= \sqrt{9^2 \times 10} = \sqrt{81 \times 10}$

$= \sqrt{810}$

**f** $3\sqrt{11}$

$= \sqrt{3^2 \times 11} = \sqrt{9 \times 11}$

$= \sqrt{99}$

## Challenge Yourself

**1** $\sqrt{432} = 12\sqrt{k}$

$\sqrt{432} = \sqrt{144 \times 3}$

$= \sqrt{144} \times \sqrt{3}$

$= 12\sqrt{3}$

$\therefore k = 3$

**2** $\sqrt{450} = c\sqrt{2}$

$\sqrt{450} = \sqrt{225 \times 2}$

$= \sqrt{225} \times \sqrt{2}$

$= 15\sqrt{2}$

$\therefore c = 15$

**3** $\sqrt{1200} = a\sqrt{b}$

$= \sqrt{400 \times 3} = \sqrt{400} \times \sqrt{3}$

$= 20\sqrt{3}$

$\therefore a = 20 \; and \; b = 3$

**4** **a** $\frac{2}{7}\sqrt{98}$

$= \frac{2}{7}\sqrt{49 \times 2}$

$= \frac{2}{7}\sqrt{49} \times \sqrt{2}$

$= \frac{2}{7} \times 7 \times \sqrt{2}$

$= 2\sqrt{2}$

**b** $\frac{1}{9}\sqrt{162}$

$= \frac{1}{9}\sqrt{81 \times 2}$

$= \frac{1}{9}\sqrt{81} \times \sqrt{2}$

$= \frac{1}{9} \times 9 \times \sqrt{2}$

$= \sqrt{2}$

**c** $\frac{3}{10}\sqrt{175}$

$= \frac{3}{10}\sqrt{25 \times 7}$

$= \frac{3}{10}\sqrt{25} \times \sqrt{7}$

$= \frac{3}{10} \times 5 \times \sqrt{7}$

$= \frac{15\sqrt{7}}{10}$

$= \frac{3\sqrt{7}}{2}$

**d** $\frac{3}{14}\sqrt{147}$

$= \frac{3}{14}\sqrt{49 \times 3}$

$= \frac{3}{14}\sqrt{49} \times \sqrt{3}$

$= \frac{3}{14} \times 7 \times \sqrt{3}$

$= \frac{21\sqrt{7}}{14}$

$= \frac{3\sqrt{3}}{2}$

# Square and Square Roots of Surds

## Multiple Choice

**1** **Option B**

$\left(\sqrt{5}\right)^2$

$= \sqrt{5} \times \sqrt{5}$

$= 5$

**2** **Option D**

$\left(2\sqrt{4x}\right)^2$

$= 2^2 \times \left(\sqrt{4x}\right)^2$

$= 4 \times 4x$

$= 16x$

**3** **Option C**

$\sqrt{81a^4}$

$= \sqrt{(9a^2)^2}$

$= 9a^2$

**4** **Option C**

$\sqrt{100x^6y^4}$

$= \sqrt{(10x^3y^2)^2}$

$= 10x^3y^2$

**5** **Option A**

$\sqrt{200x^5}$

$= \sqrt{100x^4 \times 2x}$

$= \sqrt{(10x^2)^2} \times \sqrt{2x}$

$= 10x^2\sqrt{2x}$

## Time To Practice

**1** **a** $\left(\sqrt{7}\right)^2$

$= \sqrt{7} \times \sqrt{7}$

$= 7$

**b** $\left(\sqrt{11}\right)^2$

$= \sqrt{11} \times \sqrt{11}$

$= 11$

**c** $\left(\sqrt{13}\right)^2$

$= \sqrt{13} \times \sqrt{13}$

$= 13$

**d** $\left(\sqrt{5x}\right)^2$

$= \sqrt{5x} \times \sqrt{5x}$

$= 5x$

**e** $\left(\sqrt{9x}\right)^2$

$= \sqrt{9x} \times \sqrt{9x}$

$= 9x$

**f** $\left(\sqrt{3x^2}\right)^2$

$= \sqrt{3x^2} \times \sqrt{3x^2}$

$= 3x^2$

**g** $\left(2\sqrt{2x}\right)^2$

$= 2^2 \times \left(\sqrt{2x}\right)^2$

$= 4 \times 2x$

$= 8x$

**h** $\left(-4\sqrt{3x}\right)^2$

$= (-4)^2 \times \left(\sqrt{3x}\right)^2$

$= 16 \times 3x$

$= 48x$

**i** $\left(-5\sqrt{2x^2}\right)^2$

$= (-5)^2 \times \left(\sqrt{2x^2}\right)^2$

$= 25 \times 2x^2$

$= 50x^2$

**2** **a** $\sqrt{z^2}$

$= z$

**b** $\sqrt{25z^2}$

$= \sqrt{(5z)^2}$

$= 5z$

**c** $\sqrt{121z^2}$

$= \sqrt{(11z)^2}$

$= 11z$

**d** $\sqrt{36b^2}$

$= \sqrt{(6b)^2}$

$= 6b$

**e** $\sqrt{100b^2}$

$= \sqrt{(10b)^2}$

$= 10b$

**f** $\sqrt{49b^2}$

$= \sqrt{(7b)^2}$

$= 7b$

**g** $\sqrt{144p^2q^2}$

$= \sqrt{(12pq)^2}$

$= 12pq$

**h** $\sqrt{225p^6q^4}$

$= \sqrt{(15p^3q^2)^2}$

$= 15p^3q^2$

**i** $\sqrt{289p^2q^8}$

$= \sqrt{(17pq^4)^2}$

$= 17pq^4$

**3** **a** $\sqrt{12x^2}$

$= \sqrt{4x^2 \times 3}$

$= \sqrt{(2x)^2 \times 3}$

$= \sqrt{(2x)^2} \times \sqrt{3}$

$= 2x\sqrt{3}$

**b** $\sqrt{28x^2}$

$= \sqrt{4x^2 \times 7}$

$= \sqrt{(2x)^2 \times 7}$

$= \sqrt{(2x)^2} \times \sqrt{7}$

$= 2x\sqrt{7}$

**3** **c** $\sqrt{45y^2}$

$= \sqrt{9y^2 \times 5}$
$= \sqrt{(3y)^2 \times 5}$
$= \sqrt{(3y)^2} \times \sqrt{5}$
$= 3y\sqrt{5}$

**d** $\sqrt{48x^8}$

$= \sqrt{16x^8 \times 3}$
$= \sqrt{(4x^4)^2 \times 3}$
$= \sqrt{(4x^4)^2} \times \sqrt{3}$
$= 4x^4\sqrt{3}$

**e** $\sqrt{50x^6}$

$= \sqrt{25x^6 \times 2}$
$= \sqrt{(5x^3)^2 \times 2}$
$= \sqrt{(5x^3)^2} \times \sqrt{2}$
$= 5x^3\sqrt{2}$

**f** $\sqrt{72y^4}$

$= \sqrt{36y^4 \times 2}$
$= \sqrt{(6y^2)^2 \times 2}$
$= \sqrt{(6y^2)^2} \times \sqrt{2}$
$= 6y^2\sqrt{2}$

**g** $\sqrt{75x^8y^6}$

$= \sqrt{25x^8y^6 \times 3}$
$= \sqrt{(5x^4y^3)^2 \times 3}$
$= \sqrt{(5x^4y^3)^2} \times \sqrt{3}$
$= 5x^4y^3\sqrt{3}$

**h** $\sqrt{96x^4y^2}$

$= \sqrt{16x^4y^2 \times 6}$
$= \sqrt{(4x^2y)^2 \times 6}$
$= \sqrt{(4x^2y)^2} \times \sqrt{6}$
$= 4x^2y\sqrt{6}$

**i** $\sqrt{98x^2y^6}$

$= \sqrt{49x^2y^6 \times 2}$
$= \sqrt{(7xy^3)^2 \times 2}$
$= \sqrt{(7xy^3)^2} \times \sqrt{2}$
$= 7xy^3\sqrt{2}$

**4** **a** $\sqrt{a^{13}}$

$= \sqrt{a^{12} \times a}$

$= \sqrt{(a^6)^2 \times a}$

$= \sqrt{(a^6)^2} \times \sqrt{a}$

$= a^6\sqrt{a}$

**b** $\sqrt{2b^5}$

$= \sqrt{b^4 \times 2b}$

$= \sqrt{(b^2)^2 \times 2b}$

$= \sqrt{(b^2)^2} \times \sqrt{2b}$

$= b^2\sqrt{2b}$

**c** $\sqrt{4a^7}$

$= \sqrt{4a^6 \times a}$

$= \sqrt{(2a^3)^2 \times a}$

$= \sqrt{(2a^3)^2} \times \sqrt{a}$

$= 2a^3\sqrt{a}$

**d** $3\sqrt{16a^9}$

$= 3\sqrt{16a^8 \times a}$

$= 3\sqrt{(4a^4)^2 \times a}$

$= 3\sqrt{(4a^4)^2} \times \sqrt{a}$

$= 3 \times 4a^4\sqrt{a}$

$= 12a^4\sqrt{a}$

**e** $2\sqrt{4a^3b}$

$= 2\sqrt{4a^2 \times ab}$

$= 2\sqrt{(2a)^2 \times ab}$

$= 2\sqrt{(2a)^2} \times \sqrt{ab}$

$= 2 \times 2a\sqrt{ab}$

$= 4a\sqrt{ab}$

**f** $\sqrt{45a^7b^5}$

$= \sqrt{9a^6b^4 \times 5ab}$

$= \sqrt{(3a^3b^2)^2 \times 5ab}$

$= \sqrt{(3a^3b^2)^2} \times \sqrt{5ab}$

$= 3a^3b^2\sqrt{5ab}$

4 g $3\sqrt{14ab^5}$

$= 3\sqrt{b^4 \times 14ab}$

$= 3\sqrt{(b^2)^2 \times 14ab}$

$= 3\sqrt{(b^2)^2} \times \sqrt{14ab}$

$= 3b^2\sqrt{14ab}$

h $4\sqrt{18a^{15}b^3}$

$= 4\sqrt{9a^{14}b^2 \times 2ab}$

$= 4\sqrt{(3a^7b)^2 \times 2ab}$

$= 4\sqrt{(3a^7b)^2} \times \sqrt{2ab}$

$= 4 \times 3a^7b \times \sqrt{2ab}$

$= 12a^7b\sqrt{2ab}$

i $5\sqrt{15a^5b^9}$

$= 5\sqrt{a^4b^8 \times 15ab}$

$= 5\sqrt{(a^2b^4)^2 \times 15ab}$

$= 5\sqrt{(a^2b^4)^2} \times \sqrt{15ab}$

$= 5 \times a^2b^4 \times \sqrt{15ab}$

$= 5a^2b^4\sqrt{15ab}$

## Challenge Yourself

1 a $3\sqrt{125x^4y^4}$

$= 3\sqrt{25x^4y^4 \times 5}$

$= 3\sqrt{(5x^2y^2)^2 \times 5}$

$= 3\sqrt{(5x^2y^2)^2} \times \sqrt{5}$

$= 3 \times 5x^2y^2 \times \sqrt{5}$

$= 15x^2y^2\sqrt{5}$

b $7\sqrt{162x^2y^6}$

$= 7\sqrt{81x^2y^6 \times 2}$

$= 7\sqrt{(9xy^3)^2 \times 2}$

$= 7\sqrt{(9xy^3)^2} \times \sqrt{2}$

$= 7 \times 9xy^3 \times \sqrt{2}$

$= 63xy^3\sqrt{2}$

1 c $\frac{1}{2}\sqrt{88x^4y^{12}}$

$= \frac{1}{2}\sqrt{4x^4y^{12} \times 22}$

$= \frac{1}{2}\sqrt{(2x^2y^6)^2} \times \sqrt{22}$

$= \frac{1}{2} \times 2x^2y^6 \times \sqrt{22}$

$= x^2y^6\sqrt{22}$

d $\frac{3}{5}\sqrt{80x^{18}y^{16}}$

$= \frac{3}{5}\sqrt{16x^{18}y^{16} \times 5}$

$= \frac{3}{5}\sqrt{(4x^9y^8)^2} \times \sqrt{5}$

$= \frac{3}{5} \times 4x^9y^8 \times \sqrt{5}$

$= \frac{12x^9y^8\sqrt{5}}{5}$

e $\frac{1}{15}\sqrt{75x^4y^6z^8}$

$= \frac{1}{15}\sqrt{25x^4y^6z^8 \times 3}$

$= \frac{1}{15}\sqrt{(5x^2y^3z^4)^2} \times \sqrt{3}$

$= \frac{1}{15} \times 5x^2y^3z^4 \times \sqrt{3}$

$= \frac{x^2y^3z^4\sqrt{3}}{3}$

f $\frac{2}{3}\sqrt{54x^8y^2z^4}$

$= \frac{2}{3}\sqrt{9x^8y^2z^4 \times 6}$

$= \frac{2}{3}\sqrt{(3x^4yz^2)^2} \times \sqrt{6}$

$= \frac{2}{3} \times 3x^4yz^2 \times \sqrt{6}$

$= 2x^4yz^2\sqrt{6}$

g $\frac{3}{2}\sqrt{48x^9y^3z^5}$

$= \frac{3}{2}\sqrt{16x^8y^2z^4 \times 3xyz}$

$= \frac{3}{2}\sqrt{(4x^4yz^2)^2} \times \sqrt{3xyz}$

$= \frac{3}{2} \times 4x^4yz^2 \times \sqrt{3xyz}$

$= 6x^4yz^2\sqrt{3xyz}$

h $\frac{5}{2}\sqrt{108x^7yz^5}$

$= \frac{5}{2}\sqrt{36x^6z^4 \times 3xyz}$

$= \frac{5}{2}\sqrt{(6x^3z^2)^2} \times \sqrt{3xyz}$

$= \frac{5}{2} \times 6x^3z^2 \times \sqrt{3xyz}$

$= 15x^3z^2\sqrt{3xyz}$

# Adding and Subtracting Surds

## Multiple Choice

**1** **Option A**

$3\sqrt{5} + 2\sqrt{5}$

$= 5\sqrt{5}$

**2** **Option C**

$4\sqrt{7} - \sqrt{7}$

$= 3\sqrt{7}$

**3** **Option D**

$\sqrt{12} + \sqrt{27}$

$= \sqrt{4 \times 3} + \sqrt{9 \times 3}$

$= 2\sqrt{3} + 3\sqrt{3}$

$= 5\sqrt{3}$

**4** **Option A**

$4\sqrt{50} - 5\sqrt{8}$

$= 4\sqrt{25 \times 2} - 5\sqrt{4 \times 2}$

$= 4 \times 5\sqrt{2} - 5 \times 2\sqrt{2}$

$= 20\sqrt{2} - 10\sqrt{2}$

$= 10\sqrt{2}$

**5** **Option C**

$\sqrt{18x} - 2\sqrt{8x}$

$= \sqrt{9 \times 2x} - 2\sqrt{4 \times 2x}$

$= 3\sqrt{2x} - 2 \times 2\sqrt{2x}$

$= 3\sqrt{2x} - 4\sqrt{2x}$

$= -\sqrt{2x}$

## Time To Practice

**1** **a** $4\sqrt{7} + 3\sqrt{7}$

$= (4 + 3)\sqrt{7}$
$= 7\sqrt{7}$

**b** $5\sqrt{3} + 8\sqrt{3}$

$= (5 + 8)\sqrt{3}$
$= 13\sqrt{3}$

**c** $12\sqrt{2} + 3\sqrt{2}$

$= (12 + 3)\sqrt{2}$
$= 15\sqrt{2}$

**d** $6\sqrt{3} - 2\sqrt{3}$

$= (6 - 2)\sqrt{3}$
$= 4\sqrt{3}$

**e** $11\sqrt{10} - 6\sqrt{10}$

$= (11 - 6)\sqrt{10}$
$= 5\sqrt{10}$

**f** $15\sqrt{5} - \sqrt{5}$

$= (15 - 1)\sqrt{5}$
$= 14\sqrt{5}$

**g** $-2\sqrt{13} - \sqrt{13} + 4\sqrt{13}$

$= (-2 - 1 + 4)\sqrt{13}$
$= \sqrt{13}$

**h** $-\sqrt{21} + 5\sqrt{21} - 2\sqrt{21}$

$= (-1 + 5 - 2)\sqrt{21}$
$= 2\sqrt{21}$

**2** **a** $8\sqrt{5} + 3\sqrt{3} + 7\sqrt{5} + 2\sqrt{3}$

$= (8 + 7)\sqrt{5} + (3 + 2)\sqrt{3}$
$= 15\sqrt{5} + 5\sqrt{3}$

**b** $12\sqrt{3} - 8\sqrt{7} + 5\sqrt{3} - 10\sqrt{7}$

$= (12 + 5)\sqrt{3} + (-8 - 10)\sqrt{7}$
$= 17\sqrt{3} - 18\sqrt{7}$

**c** $2\sqrt{2} - 4\sqrt{10} - 5\sqrt{2} + \sqrt{10}$

$= (2 - 5)\sqrt{2} + (-4 + 1)\sqrt{10}$
$= -3\sqrt{2} - 3\sqrt{10}$

**d** $-4\sqrt{11} - 2\sqrt{15} + 7\sqrt{11} + 4\sqrt{15}$

$= (-4 + 7)\sqrt{11} + (-2 + 4)\sqrt{15}$
$= 3\sqrt{11} + 2\sqrt{15}$

**3** **a** $\sqrt{8} + 3\sqrt{2}$

$= \sqrt{4 \times 2} + 3\sqrt{2}$

$= 2\sqrt{2} + 3\sqrt{2}$

$= 5\sqrt{2}$

**b** $\sqrt{20} - \sqrt{5}$

$= \sqrt{4 \times 5} - \sqrt{5}$

$= 2\sqrt{5} - \sqrt{5}$

$= \sqrt{5}$

**c** $2\sqrt{27} + 4\sqrt{3}$

$= 2\sqrt{9 \times 3} + 4\sqrt{3}$

$= 2 \times 3\sqrt{3} + 4\sqrt{3}$

$= 6\sqrt{3} + 4\sqrt{3}$

$= 10\sqrt{3}$

**d** $2\sqrt{24} + 3\sqrt{6}$

$= 2\sqrt{4 \times 6} + 3\sqrt{6}$

$= 2 \times 2\sqrt{6} + 3\sqrt{6}$

$= 4\sqrt{6} + 3\sqrt{6}$

$= 7\sqrt{6}$

**e** $3\sqrt{72} + 2\sqrt{98}$

$= 3\sqrt{36 \times 2} + 3\sqrt{49 \times 2}$

$= 3 \times 6\sqrt{2} + 2 \times 7\sqrt{2}$

$= 18\sqrt{2} + 14\sqrt{2}$

$= 32\sqrt{2}$

**f** $2\sqrt{125} - 3\sqrt{45}$

$= 2\sqrt{25 \times 5} - 3\sqrt{9 \times 5}$

$= 2 \times 5\sqrt{5} - 3 \times 3\sqrt{5}$

$= 10\sqrt{5} - 9\sqrt{5}$

$= \sqrt{5}$

**g** $2\sqrt{80} - \sqrt{45} + 2\sqrt{63}$

$= 2\sqrt{16 \times 5} - \sqrt{9 \times 5} + 2\sqrt{9 \times 7}$

$= 2 \times 4\sqrt{5} - 3\sqrt{5} + 2 \times 3\sqrt{7}$

$= 8\sqrt{5} - 3\sqrt{5} + 6\sqrt{7}$

$= 5\sqrt{5} + 6\sqrt{7}$

**h** $\sqrt{36} - \sqrt{108} - 3\sqrt{3}$

$= 6 - \sqrt{36 \times 3} - 3\sqrt{3}$

$= 6 - 6\sqrt{3} - 3\sqrt{3}$

$= 6 - 9\sqrt{3}$

**4** **a** $2\sqrt{x} + 5\sqrt{y} + 6\sqrt{x} - 2\sqrt{y}$

$= 8\sqrt{x} + 3\sqrt{y}$

**b** $7\sqrt{x} - \sqrt{y} + 2\sqrt{9x} - 3\sqrt{y}$

$= 7\sqrt{x} - 4\sqrt{y} + 2 \times 3\sqrt{x}$

$= 7\sqrt{x} - 4\sqrt{y} + 6\sqrt{x}$

$= 13\sqrt{x} - 4\sqrt{y}$

**c** $10\sqrt{y} - 2\sqrt{4y} + 2\sqrt{x} - \sqrt{x}$

$= 10\sqrt{y} - 2 \times 2\sqrt{y} + \sqrt{x}$

$= 10\sqrt{y} - 4\sqrt{y} + \sqrt{x}$

$= 6\sqrt{y} + \sqrt{x}$

**d** $6\sqrt{x} - 12\sqrt{9y} - 7\sqrt{x} + 2\sqrt{y}$

$= -\sqrt{x} - 12 \times 3\sqrt{y} + 2\sqrt{y}$

$= -\sqrt{x} - 36\sqrt{y} + 2\sqrt{y}$

$= -\sqrt{x} - 34\sqrt{y}$

## Challenge Yourself

**1** $4\sqrt{2} + \sqrt{32}$

$= 4\sqrt{2} + \sqrt{16 \times 2}$

$= 4\sqrt{2} + 4\sqrt{2}$

$= 8\sqrt{2}$

$\therefore a = 8 \text{ and } b = 2$

**2** $3\sqrt{125} - 2\sqrt{45}$

$= 3\sqrt{25 \times 5} - 2\sqrt{9 \times 5}$

$= 3 \times 5\sqrt{5} - 2 \times 3\sqrt{5}$

$= 15\sqrt{5} - 6\sqrt{5}$

$= 9\sqrt{5}$

$\therefore a = 9 \text{ and } b = 5$

**3** $4\sqrt{50x} - 3\sqrt{18x}$

$= 4\sqrt{25 \times 2x} - 3\sqrt{9 \times 2x}$

$= 4 \times 5\sqrt{2x} - 3 \times 3\sqrt{2x}$

$= 20\sqrt{2x} - 9\sqrt{2x}$

$= 11\sqrt{2x}$

$\therefore a = 11 \text{ and } b = 2$

**4** **a** $10\sqrt{3a} - 15\sqrt{27a} + 8\sqrt{12a}$

$= 10\sqrt{3a} - 15\sqrt{9 \times 3a} + 8\sqrt{4 \times 3a}$

$= 10\sqrt{3a} - 15 \times 3\sqrt{3a} + 8 \times 2\sqrt{3a}$

$= 10\sqrt{3a} - 45\sqrt{3a} + 16\sqrt{3a}$

$= -19\sqrt{3a}$

**b** $6\sqrt{2b} + \sqrt{128b} - \sqrt{144}$

$= 6\sqrt{2b} + \sqrt{64 \times 2b} - 12$

$= 6\sqrt{2b} + 8\sqrt{2b} - 12$

$= 14\sqrt{2b} - 12$

**c** $\frac{1}{8}\sqrt{32a} - \frac{7}{6}\sqrt{18a} - \frac{1}{3}\sqrt{72a}$

$= \frac{1}{8}\sqrt{16 \times 2a} - \frac{7}{6}\sqrt{9 \times 2a} - \frac{1}{3}\sqrt{36 \times 2a}$

$= \frac{1}{8} \times 4\sqrt{2a} - \frac{7}{6} \times 3\sqrt{2a} - \frac{1}{3} \times 6\sqrt{2a}$

$= \frac{1}{2}\sqrt{2a} - \frac{7}{2}\sqrt{2a} - 2\sqrt{2a}$

$= -5\sqrt{2a}$

**d** $\frac{1}{2}\sqrt{150ab} + \frac{1}{4}\sqrt{96ab} - \frac{1}{6}\sqrt{6ab}$

$= \frac{1}{2}\sqrt{25 \times 6ab} + \frac{1}{4}\sqrt{16 \times 6ab} - \frac{1}{6}\sqrt{6ab}$

$= \frac{1}{2} \times 5\sqrt{6ab} + \frac{1}{4} \times 4\sqrt{6ab} - \frac{1}{6}\sqrt{6ab}$

$= \frac{5}{2}\sqrt{6ab} + \sqrt{6ab} - \frac{1}{6}\sqrt{6ab}$

$= \frac{10\sqrt{6ab}}{3}$

# Multiplying Surds

## Multiple Choice

1 **Option B**

$\sqrt{5} \times \sqrt{2}$

$= \sqrt{5 \times 2}$

$= \sqrt{10}$

2 **Option A**

$2\sqrt{18} \times \sqrt{3}$

$= 2\sqrt{54}$

$= 2\sqrt{9 \times 6}$

$= 2 \times 3 \times \sqrt{6}$

$= 6\sqrt{6}$

3 **Option A**

$3\sqrt{5} \times 2\sqrt{10}$

$= 6\sqrt{5 \times 10}$

$= 6\sqrt{50}$

$= 6\sqrt{25 \times 2}$

$= 6 \times 5\sqrt{2}$

$= 30\sqrt{2}$

4 **Option C**

$2\sqrt{45} \times \sqrt{63}$

$= 2\sqrt{9 \times 5} \times \sqrt{9 \times 7}$

$= 2 \times 3\sqrt{5} \times 3\sqrt{7}$

$= 18\sqrt{5 \times 7}$

$= 18\sqrt{35}$

5 **Option C**

$\sqrt{12x^2} \times \sqrt{3}$

$= \sqrt{12x^2 \times 3}$

$= \sqrt{36x^2}$

$= 6x$

## Time To Practice

**1** **a** $\sqrt{2} \times \sqrt{7}$

$= \sqrt{2 \times 7}$
$= \sqrt{14}$

**b** $\sqrt{6} \times \sqrt{7}$

$= \sqrt{6 \times 7}$
$= \sqrt{42}$

**c** $\sqrt{5} \times \sqrt{6}$

$= \sqrt{5 \times 6}$
$= \sqrt{30}$

**d** $\sqrt{10} \times \sqrt{11}$

$= \sqrt{10 \times 11}$
$= \sqrt{110}$

**e** $\sqrt{2} \times \sqrt{17}$

$= \sqrt{2 \times 17}$
$= \sqrt{34}$

**f** $3\sqrt{2} \times \sqrt{21}$

$= 3\sqrt{2 \times 21}$
$= 3\sqrt{42}$

**g** $2\sqrt{11} \times 5\sqrt{3}$

$= 10\sqrt{11 \times 3}$
$= 10\sqrt{33}$

**h** $6\sqrt{3} \times 10\sqrt{7}$

$= 60\sqrt{3 \times 7}$
$= 60\sqrt{21}$

**i** $3\sqrt{5} \times 4\sqrt{7}$

$= 12\sqrt{5 \times 7}$
$= 12\sqrt{35}$

**2** **a** $2\sqrt{8} \times \sqrt{2}$

$= 2\sqrt{8 \times 2}$
$= 2\sqrt{16}$
$= 2 \times 4$
$= 8$

**b** $3\sqrt{18} \times \sqrt{12}$

$= 3\sqrt{9 \times 2} \times \sqrt{4 \times 3}$
$= 3 \times 3\sqrt{2} \times 2\sqrt{3}$
$= 18\sqrt{2 \times 3}$
$= 18\sqrt{6}$

**2** **c** $\sqrt{12} \times 4\sqrt{48}$

$= \sqrt{4 \times 3} \times 4\sqrt{16 \times 3}$

$= 2\sqrt{3} \times 4 \times 4\sqrt{3}$

$= 2\sqrt{3} \times 16\sqrt{3}$

$= 32 \times 3$

$= 96$

**d** $\sqrt{32} \times \sqrt{72}$

$= \sqrt{16 \times 2} \times \sqrt{36 \times 2}$

$= 4\sqrt{2} \times 6\sqrt{2}$

$= 24 \times 2$

$= 48$

**e** $2\sqrt{75} \times 3\sqrt{45}$

$= 2\sqrt{25 \times 3} \times 3\sqrt{9 \times 5}$

$= 2 \times 5\sqrt{3} \times 3 \times 3\sqrt{5}$

$= 10\sqrt{3} \times 9\sqrt{5}$

$= 90\sqrt{15}$

**f** $\sqrt{20} \times \sqrt{125}$

$= \sqrt{4 \times 5} \times \sqrt{25 \times 5}$

$= 2\sqrt{5} \times 5\sqrt{5}$

$= 10 \times 5$

$= 50$

**g** $4\sqrt{63} \times 2\sqrt{6}$

$= 4\sqrt{9 \times 7} \times 2\sqrt{6}$

$= 4 \times 3\sqrt{7} \times 2\sqrt{6}$

$= 24\sqrt{7 \times 6}$

$= 24\sqrt{42}$

**h** $3\sqrt{50} \times 2\sqrt{20}$

$= 3\sqrt{25 \times 2} \times 2\sqrt{4 \times 5}$

$= 3 \times 5\sqrt{2} \times 2 \times 2\sqrt{5}$

$= 15\sqrt{2} \times 4\sqrt{5}$

$= 60\sqrt{2 \times 5}$

$= 60\sqrt{10}$

**i** $4\sqrt{15} \times 2\sqrt{18}$

$= 4\sqrt{15} \times 2 \times \sqrt{9 \times 2}$

$= 4\sqrt{15} \times 2 \times 3\sqrt{2}$

$= 4\sqrt{15} \times 6\sqrt{2}$

$= 24\sqrt{15 \times 2}$

$= 24\sqrt{30}$

**3** **a** $\frac{\sqrt{5}}{3}+\frac{\sqrt{5}}{4}$

$=\frac{4\sqrt{5}+3\sqrt{5}}{12}$

$=\frac{7\sqrt{5}}{12}$

**b** $\frac{\sqrt{2}}{5}-\frac{\sqrt{2}}{6}$

$=\frac{6\sqrt{2}-5\sqrt{2}}{30}$

$=\frac{\sqrt{2}}{30}$

**c** $\frac{\sqrt{7}}{4}-\frac{\sqrt{7}}{12}$

$=\frac{3\sqrt{7}}{12}-\frac{\sqrt{7}}{12}$

$=\frac{2\sqrt{7}}{12}$

$=\frac{\sqrt{7}}{6}$

**d** $\frac{2\sqrt{2}}{5}-\frac{\sqrt{2}}{2}$

$=\frac{4\sqrt{2}-5\sqrt{2}}{10}$

$=-\frac{\sqrt{2}}{10}$

**e** $\frac{3\sqrt{3}}{7}-\frac{\sqrt{3}}{2}$

$=\frac{6\sqrt{3}-7\sqrt{3}}{14}$

$=-\frac{\sqrt{3}}{14}$

**f** $\frac{7\sqrt{7}}{4}+\frac{\sqrt{7}}{12}$

$=\frac{21\sqrt{7}}{12}+\frac{\sqrt{7}}{12}$

$=\frac{22\sqrt{7}}{12}$

$=\frac{11\sqrt{7}}{6}$

**3** **g** $\dfrac{\sqrt{8}}{3} - \dfrac{\sqrt{2}}{5}$

$= \dfrac{2\sqrt{2}}{3} - \dfrac{\sqrt{2}}{5}$

$= \dfrac{10\sqrt{2} - 3\sqrt{2}}{15}$

$= \dfrac{7\sqrt{2}}{15}$

**h** $\dfrac{\sqrt{12}}{4} + \dfrac{\sqrt{3}}{6}$

$= \dfrac{2\sqrt{3}}{4} + \dfrac{\sqrt{3}}{6}$

$= \dfrac{12\sqrt{3} + 4\sqrt{3}}{24}$

$= \dfrac{16\sqrt{3}}{24}$

$= \dfrac{2\sqrt{3}}{3}$

**i** $\dfrac{2\sqrt{18}}{3} - \dfrac{\sqrt{72}}{2}$

$= \dfrac{2 \times 3\sqrt{2}}{3} - \dfrac{6\sqrt{2}}{2}$

$= \dfrac{6\sqrt{2}}{3} - \dfrac{6\sqrt{2}}{2}$

$= \dfrac{12\sqrt{2} - 18\sqrt{2}}{6}$

$= \dfrac{-6\sqrt{2}}{6}$

$= -\sqrt{2}$

**4** **a** $\sqrt{2x^2} \times \sqrt{3}$

$= \sqrt{2x^2 \times 3}$

$= \sqrt{6x^2}$

$= x\sqrt{6}$

**b** $\sqrt{8x^2y} \times \sqrt{y}$

$= \sqrt{8x^2y \times y} = \sqrt{8x^2y^2}$

$= \sqrt{4x^2y^2 \times 2}$

$= \sqrt{(2xy)^2 \times 2}$

$= 2xy\sqrt{2}$

**4** **c** $\sqrt{2x^2y^2} \times 3\sqrt{27}$

$= xy\sqrt{2} \times 3\sqrt{9 \times 3}$

$= xy\sqrt{2} \times 3 \times 3\sqrt{3}$

$= xy\sqrt{2} \times 9\sqrt{3}$

$= 9xy\sqrt{6}$

**d** $\sqrt{12x^8y} \times 6\sqrt{2y}$

$= 6\sqrt{12x^8y \times 2y}$

$= 6\sqrt{24x^8y^2}$

$= 6\sqrt{4x^8y^2 \times 6}$

$= 6\sqrt{(2x^4y)^2 \times 6}$

$= 6 \times 2x^4y\sqrt{6}$

$= 12x^4y\sqrt{6}$

**e** $\sqrt{2x^6y^2} \times \sqrt{24x^4y^8}$

$= \sqrt{2x^6y^2 \times 24x^4y^8}$

$= \sqrt{48x^{10}y^{10}}$

$= \sqrt{16x^{10}y^{10} \times 3}$

$= \sqrt{(4x^5y^5)^2 \times 3}$

$= 4x^5y^5\sqrt{3}$

**f** $3\sqrt{7x^2y^3} \times 2\sqrt{x^2y^5}$

$= 6\sqrt{7x^2y^3 \times x^2y^5}$

$= 6\sqrt{7x^4y^8}$

$= 6\sqrt{7 \times x^4y^8}$

$= 6\sqrt{7 \times (x^2y^4)^2}$

$= 6x^2y^4\sqrt{7}$

## Challenge Yourself

**1** **a** $\frac{1}{2}\sqrt{60x^4} \times \sqrt{18x^2}$

$= \frac{1}{2}\sqrt{4x^4 \times 15} \times \sqrt{9x^2 \times 2}$

$= \frac{1}{2}\sqrt{(2x^2)^2 \times 15} \times \sqrt{(3x)^2 \times 2}$

$= \frac{1}{2} \times 2x^2\sqrt{15} \times 3x\sqrt{2}$

$= 3x^3\sqrt{15 \times 2}$

$= 3x^3\sqrt{30}$

**b** $\frac{1}{4}\sqrt{128x^8} \times 2\sqrt{75x^2}$

$= \frac{1}{4}\sqrt{64x^8 \times 2} \times 2\sqrt{25x^2 \times 3}$

$= \frac{1}{4}\sqrt{(8x^4)^2 \times 2} \times 2\sqrt{(5x)^2 \times 3}$

$= \frac{1}{4} \times 8x^4\sqrt{2} \times 2 \times 5x\sqrt{3}$

$= 2x^4\sqrt{2} \times 10x\sqrt{3}$

$= 20x^5\sqrt{6}$

1 c $\frac{2}{3}\sqrt{12x^8y^2} \times 6\sqrt{32y^6}$

$= \frac{2}{3}\sqrt{4x^8y^2 \times 3} \times 6\sqrt{16y^6 \times 2}$

$= \frac{2}{3}\sqrt{(2x^4y)^2 \times 3} \times 6\sqrt{(4y^3)^2 \times 2}$

$= \frac{2}{3} \times 2x^4y\sqrt{3} \times 6 \times 4y^3\sqrt{2}$

$= \frac{4x^4y\sqrt{3}}{3} \times 24y^3\sqrt{2}$

$= 4x^4y\sqrt{3} \times 8y^3\sqrt{2}$

$= 32x^4y^4\sqrt{3 \times 2}$

$= 32x^4y^4\sqrt{6}$

d $\frac{3}{8}\sqrt{63x^2y^2} \times \frac{6}{7}\sqrt{96y^{12}}$

$= \frac{3}{8}\sqrt{9x^2y^2 \times 7} \times \frac{6}{7}\sqrt{16y^{12} \times 6}$

$= \frac{3}{8}\sqrt{(3xy)^2 \times 7} \times \frac{6}{7}\sqrt{(4y^6)^2 \times 6}$

$= \frac{3}{8} \times 3xy\sqrt{7} \times \frac{6}{7} \times 4y^6\sqrt{6}$

$= \frac{9xy\sqrt{7}}{8} \times \frac{24y^6\sqrt{6}}{7}$

$= \frac{9xy\sqrt{7}}{1} \times \frac{3y^6\sqrt{6}}{7}$

$= \frac{27xy^7\sqrt{7 \times 6}}{7}$

$= \frac{27xy^7\sqrt{42}}{7}$

**1** **e** $\frac{2}{3}\sqrt{88x^4y^2} \times \frac{4}{5}\sqrt{48y^6}$

$= \frac{2}{3}\sqrt{4x^4y^2 \times 22} \times \frac{4}{5}\sqrt{16y^6 \times 3}$

$= \frac{2}{3}\sqrt{(2x^2y)^2 \times 22} \times \frac{4}{5}\sqrt{(4y^3)^2 \times 3}$

$= \frac{2}{3} \times 2x^2y\sqrt{22} \times \frac{4}{5} \times 4y^3\sqrt{3}$

$= \frac{4x^2y\sqrt{22}}{3} \times \frac{16y^3\sqrt{3}}{5}$

$= \frac{64x^2y^4\sqrt{22 \times 3}}{15}$

$= \frac{64x^2y^4\sqrt{66}}{15}$

**f** $\frac{1}{6}\sqrt{200x^6y^4} \times \frac{3}{7}\sqrt{60x^8y^2}$

$= \frac{1}{6}\sqrt{100x^6y^4 \times 2} \times \frac{3}{7}\sqrt{4x^8y^2 \times 15}$

$= \frac{1}{6}\sqrt{(10x^3y^2)^2 \times 2} \times \frac{3}{7}\sqrt{(2x^4y)^2 \times 15}$

$= \frac{1}{6} \times 10x^3y^2\sqrt{2} \times \frac{3}{7} \times 2x^4y\sqrt{15}$

$= \frac{10x^3y^2\sqrt{2}}{6} \times \frac{6x^4y\sqrt{15}}{7}$

$= \frac{10x^7y^3\sqrt{2 \times 15}}{7}$

$= \frac{60x^7y^3\sqrt{30}}{7}$

# Dividing Surds

## Multiple Choice

**1** **Option C**

$\sqrt{48} \div \sqrt{12}$

$= \sqrt{\dfrac{48}{12}}$

$= \sqrt{4}$

$= 2$

**2** **Option A**

$\sqrt{18} \div \sqrt{2}$

$= \sqrt{\dfrac{18}{2}}$

$= \sqrt{9}$

$= 3$

**3** **Option B**

$\sqrt{75} \div \sqrt{5}$

$= \sqrt{\dfrac{75}{5}}$

$= \sqrt{15}$

**4** **Option C**

$3\sqrt{27} \div \sqrt{3}$

$= 3\sqrt{\dfrac{27}{9}}$

$= 3\sqrt{3}$

**5** **Option D**

$\sqrt{72x^2} \div \sqrt{8}$

$= \sqrt{\dfrac{72x^2}{8}}$

$= \sqrt{9x^2}$

$= \sqrt{(3x)^2}$

$= 3x$

## Time To Practice

**1** **a** $\sqrt{16} \div \sqrt{4}$

$= \sqrt{\frac{16}{4}}$

$= \sqrt{4} = 2$

**b** $\sqrt{25} \div \sqrt{5}$

$= \sqrt{\frac{25}{5}}$

$= \sqrt{5}$

**c** $\sqrt{81} \div \sqrt{9}$

$= \sqrt{\frac{81}{9}}$

$= \sqrt{9}$

$= 3$

**d** $\sqrt{64} \div \sqrt{2}$

$= \sqrt{\frac{64}{2}} = \sqrt{32}$

$= \sqrt{16 \times 2}$

$= 4\sqrt{2}$

**e** $3\sqrt{36} \div 2\sqrt{4}$

$= \frac{3}{2}\sqrt{\frac{36}{4}}$

$= \frac{3}{2}\sqrt{9}$

$= \frac{3}{2} \times 3$

$= \frac{9}{2}$

**f** $2\sqrt{45} \div \sqrt{5}$

$= 2\sqrt{\frac{45}{5}}$

$= 2\sqrt{9}$

$= 2 \times 3$

$= 6$

**g** $4\sqrt{72} \div 2\sqrt{18}$

$= \frac{4}{2}\sqrt{\frac{72}{18}}$

$= 2\sqrt{4}$

$= 2 \times 2$

$= 4$

**h** $\sqrt{200} \div 2\sqrt{5}$

$= \frac{1}{2}\sqrt{\frac{200}{5}}$

$= \frac{1}{2}\sqrt{40} = \frac{1}{2}\sqrt{4 \times 10}$

$= \frac{1}{2} \times 2\sqrt{10}$

$= \sqrt{10}$

**1** **i** $8\sqrt{18} \div 4\sqrt{2}$

$= \frac{8}{4}\sqrt{\frac{18}{2}}$

$= 2\sqrt{9}$

$= 2 \times 3$

$= 6$

**j** $\sqrt{\frac{9}{64}}$

$= \frac{\sqrt{9}}{\sqrt{64}}$

$= \frac{3}{8}$

**k** $\sqrt{\frac{12}{25}}$

$= \frac{\sqrt{12}}{\sqrt{25}}$

$= \frac{\sqrt{4 \times 3}}{5}$

$= \frac{2\sqrt{3}}{5}$

**l** $3\sqrt{\frac{27}{49}}$

$= \frac{3\sqrt{27}}{\sqrt{49}}$

$= \frac{3\sqrt{9 \times 3}}{7}$

$= \frac{3 \times 3\sqrt{3}}{7}$

$= \frac{9\sqrt{3}}{7}$

**2** **a** $\frac{3\sqrt{64}}{5\sqrt{8}}$

$= \frac{3}{5}\sqrt{\frac{64}{8}} = \frac{3}{5}\sqrt{8}$

$= \frac{3}{5}\sqrt{4 \times 3}$

$= \frac{3}{5} \times 2\sqrt{2}$

$= \frac{6\sqrt{2}}{5}$

**b** $\frac{\sqrt{60}}{5\sqrt{5}}$

$= \frac{1}{5}\sqrt{\frac{60}{5}} = \frac{1}{5}\sqrt{12}$

$= \frac{1}{5}\sqrt{4 \times 3}$

$= \frac{1}{5} \times 2\sqrt{3}$

$= \frac{2\sqrt{3}}{5}$

**2** **c** $\dfrac{9\sqrt{80}}{12\sqrt{4}}$

$= \dfrac{3}{4}\sqrt{\dfrac{80}{4}} = \dfrac{3}{4}\sqrt{20}$

$= \dfrac{3}{4}\sqrt{4 \times 5}$

$= \dfrac{3}{4} \times 2\sqrt{5}$

$= \dfrac{3\sqrt{5}}{2}$

**d** $\dfrac{3\sqrt{88}}{14\sqrt{11}}$

$= \dfrac{3}{14}\sqrt{\dfrac{88}{11}} = \dfrac{3}{14}\sqrt{8}$

$= \dfrac{3}{14}\sqrt{4 \times 2}$

$= \dfrac{3}{14} \times 2\sqrt{2}$

$= \dfrac{3\sqrt{2}}{7}$

**e** $\dfrac{3\sqrt{108}}{11\sqrt{6}}$

$= \dfrac{3}{11}\sqrt{\dfrac{108}{6}} = \dfrac{3}{11}\sqrt{18}$

$= \dfrac{3}{11}\sqrt{9 \times 2}$

$= \dfrac{3}{11} \times 3\sqrt{2}$

$= \dfrac{9\sqrt{2}}{11}$

**f** $\dfrac{12\sqrt{56}}{18\sqrt{2}}$

$= \dfrac{2}{3}\sqrt{\dfrac{56}{2}} = \dfrac{2}{3}\sqrt{28}$

$= \dfrac{2}{3}\sqrt{4 \times 7}$

$= \dfrac{2}{3} \times 2\sqrt{7}$

$= \dfrac{4\sqrt{7}}{3}$

**3** **a** $\frac{1}{\sqrt{2}} \times \frac{\sqrt{6}}{3}$

$= \frac{\sqrt{6}}{3\sqrt{2}}$

$= \frac{1}{3}\sqrt{\frac{6}{2}}$

$= \frac{1}{3}\sqrt{3}$

$= \frac{\sqrt{3}}{3}$

**b** $\frac{\sqrt{10}}{\sqrt{3}} \times \frac{\sqrt{6}}{2}$

$= \frac{\sqrt{60}}{2\sqrt{3}}$

$= \frac{1}{2}\sqrt{\frac{60}{3}} = \frac{1}{2}\sqrt{20}$

$= \frac{1}{2}\sqrt{4 \times 5}$

$= \frac{1}{2} \times 2\sqrt{5}$

$= \sqrt{5}$

**c** $\frac{\sqrt{12}}{3} \times \frac{\sqrt{9}}{2\sqrt{6}}$

$= \frac{\sqrt{108}}{6\sqrt{6}}$

$= \frac{1}{6}\sqrt{\frac{108}{6}} = \frac{1}{6}\sqrt{18}$

$= \frac{1}{6}\sqrt{9 \times 2}$

$= \frac{1}{6} \times 3\sqrt{2}$

$= \frac{\sqrt{2}}{2}$

**d** $\frac{\sqrt{5}}{2} \div \frac{4}{\sqrt{10}}$

$= \frac{\sqrt{5}}{2} \times \frac{\sqrt{10}}{4}$

$= \frac{\sqrt{50}}{8}$

$= \frac{\sqrt{25 \times 2}}{8}$

$= \frac{5\sqrt{2}}{8}$

**3** **e** $\frac{\sqrt{18}}{\sqrt{2}} \div \frac{3}{\sqrt{6}}$

$= \sqrt{\frac{18}{2}} \times \frac{\sqrt{6}}{3}$

$= \sqrt{9} \times \frac{\sqrt{6}}{3}$

$= 3 \times \frac{\sqrt{6}}{3}$

$= \sqrt{6}$

**f** $\frac{5}{\sqrt{6}} \div \frac{2}{9\sqrt{48}}$

$= \frac{5}{\sqrt{6}} \times \frac{9\sqrt{48}}{2}$

$= \frac{45\sqrt{48}}{2\sqrt{6}}$

$= \frac{45}{2}\sqrt{\frac{48}{6}}$

$= \frac{45}{2}\sqrt{8}$

$= \frac{45}{2}\sqrt{4 \times 2}$

$= \frac{45}{2} \times 2\sqrt{2}$

$= 45\sqrt{2}$

**4** **a** $\frac{\sqrt{x^8y^4}}{\sqrt{x^2y^2}}$

$= \frac{\sqrt{(x^4y^2)^2}}{\sqrt{(xy)^2}}$

$= \frac{x^4y^2}{xy}$

$= x^3y$

**b** $\frac{\sqrt{8x^8y^4}}{\sqrt{2x^4y^6}}$

$= \sqrt{\frac{8x^8y^4}{2x^4y^6}}$

$= \sqrt{\frac{4x^4}{y^2}} = \sqrt{\left(\frac{2x^2}{y}\right)^2}$

$= \frac{2x^2}{y}$

**4** **c** $\dfrac{\sqrt{9x^2}}{\sqrt{x^4y^8}}$

$= \dfrac{\sqrt{(3x)^2}}{\sqrt{(x^2y^4)^2}}$

$= \dfrac{3x}{x^2y^4}$

$= \dfrac{3}{xy^4}$

**d** $\dfrac{\sqrt{x^4y^2}}{\sqrt{x^6y^4}} \div \dfrac{\sqrt{x^8}}{\sqrt{y^6}}$

$= \dfrac{\sqrt{(x^2y)^2}}{\sqrt{(x^3y^2)^2}} \div \dfrac{\sqrt{(x^4)^2}}{\sqrt{(y^3)^2}}$

$= \dfrac{x^2y}{x^3y^2} \div \dfrac{x^4}{y^3}$

$= \dfrac{1}{xy} \times \dfrac{y^3}{x^4}$

$= \dfrac{1}{x} \times \dfrac{y^2}{x^4} = \dfrac{y^2}{x^5}$

**e** $\dfrac{\sqrt{xy}}{\sqrt{x^5y^7}} \div \dfrac{\sqrt{x^2y^4}}{\sqrt{12x^8y^{12}}}$

$= \sqrt{\dfrac{xy}{x^5y^7}} \div \dfrac{\sqrt{(xy^2)^2}}{\sqrt{4x^8y^{12} \times 3}}$

$= \sqrt{\dfrac{1}{x^4y^6}} \div \dfrac{xy^2}{\sqrt{(2x^4y^6)^2 \times 3}}$

$= \dfrac{1}{\sqrt{(x^2y^3)^2}} \div \dfrac{xy^2}{2x^4y^6\sqrt{3}}$

$= \dfrac{1}{x^2y^3} \times \dfrac{2x^4y^6\sqrt{3}}{xy^2}$

$= 2xy\sqrt{3}$

**f** $\dfrac{2\sqrt{6x^5y^3}}{\sqrt{x^3y}} \div \dfrac{3\sqrt{y}}{\sqrt{y^7}}$

$= 2\sqrt{\dfrac{6x^5y^3}{x^3y}} \div 3\sqrt{\dfrac{y}{y^7}}$

$= 2\sqrt{6x^2y^2} \div 3\sqrt{\dfrac{1}{y^6}}$

$= 2\sqrt{6(xy)^2} \div 3 \times \dfrac{1}{\sqrt{(y^3)^2}}$

$= 2xy\sqrt{6} \div \dfrac{3}{y^3}$

$= 2xy\sqrt{6} \times \dfrac{y^3}{3}$

$= \dfrac{2xy^4\sqrt{6}}{3}$

## Challenge Yourself

**1** **a** $\dfrac{9\sqrt{12}\times 2\sqrt{18}}{3\sqrt{8}}$

$=\dfrac{9\sqrt{4\times 3}\times 2\sqrt{9\times 2}}{3\sqrt{4\times 2}}$

$=\dfrac{18\sqrt{3}\times 6\sqrt{2}}{6\sqrt{2}}$

$=18\sqrt{3}$

**b** $\left(\dfrac{(5\sqrt{3})^2\times 2\sqrt{5}}{3}\right)\times\dfrac{\sqrt{3}}{6\sqrt{5}}$

$=\dfrac{25\times 3\times 2\sqrt{5}}{3}\times\dfrac{\sqrt{3}}{6\sqrt{5}}$

$=\dfrac{50\sqrt{5}}{1}\times\dfrac{\sqrt{3}}{6\sqrt{5}}$

$=\dfrac{25}{1}\times\dfrac{\sqrt{3}}{3}$

$=\dfrac{25\sqrt{3}}{3}$

**c** $\dfrac{16}{\sqrt{3}}\div\left(\dfrac{(2\sqrt{2})^2\times(2\sqrt{5})^2}{5\sqrt{60}}\right)$

$=\dfrac{16}{\sqrt{3}}\div\left(\dfrac{4\times 2\times 4\times 5}{5\sqrt{4\times 15}}\right)$

$=\dfrac{16}{\sqrt{3}}\div\dfrac{160}{10\sqrt{15}}$

$=\dfrac{16}{\sqrt{3}}\times\dfrac{10\sqrt{15}}{160}$

$=\dfrac{1}{\sqrt{3}}\times\dfrac{\sqrt{15}}{1}$

$=\dfrac{1}{1}\times\dfrac{\sqrt{15}}{\sqrt{3}}$

$=\sqrt{\dfrac{15}{3}}$

$=\sqrt{5}$

**d** $\left(\dfrac{\sqrt{54}\times 3\sqrt{22}}{3\sqrt{8}}\right)\div\dfrac{16\sqrt{33}}{\sqrt{3}}$

$=\left(\dfrac{\sqrt{9\times 6}\times 3\sqrt{22}}{3\sqrt{4\times 2}}\right)\times\dfrac{\sqrt{3}}{16\sqrt{33}}$

$=\left(\dfrac{3\sqrt{6}\times 3\sqrt{22}}{6\sqrt{2}}\right)\times\dfrac{1}{16}\sqrt{\dfrac{3}{33}}$

$=\left(\dfrac{9\sqrt{6\times 22}}{6\sqrt{2}}\right)\times\dfrac{1}{16}\sqrt{\dfrac{1}{11}}$

$=\left(\dfrac{3}{2}\sqrt{\dfrac{132}{2}}\right)\times\dfrac{1}{16\sqrt{11}}$

$=\dfrac{3\sqrt{66}}{2}\times\dfrac{1}{16\sqrt{11}}$

$=\dfrac{3\sqrt{66}}{32\sqrt{11}}=\dfrac{3}{32}\sqrt{\dfrac{66}{11}}$

$=\dfrac{3\sqrt{6}}{32}$

**2** **a** $\dfrac{\sqrt{x^8y^{12}}}{\sqrt{3x^2y^3}} \div \dfrac{4\sqrt{x^5y^7}}{\sqrt{21x}}$

$= \dfrac{\sqrt{x^8y^{12}}}{\sqrt{3x^2y^3}} \times \dfrac{\sqrt{21x}}{4\sqrt{x^5y^7}}$

$= \dfrac{\sqrt{x^8y^{12} \times 21x}}{4\sqrt{3x^2y^3 \times x^5y^7}}$

$= \dfrac{\sqrt{21x^9y^{12}}}{4\sqrt{3x^7y^{10}}}$

$= \dfrac{1}{4}\sqrt{\dfrac{21x^9y^{12}}{3x^7y^{10}}}$

$= \dfrac{1}{4}\sqrt{7x^2y^2}$

$= \dfrac{1}{4}\sqrt{7(xy)^2}$

$= \dfrac{xy\sqrt{7}}{4}$

**b** $\dfrac{2\sqrt{5x^4y^2}}{\sqrt{4x^3y^5}} \div \dfrac{3\sqrt{10x^7y}}{\sqrt{6y^6}}$

$= \dfrac{2\sqrt{5x^4y^2}}{\sqrt{4x^3y^5}} \times \dfrac{\sqrt{6y^6}}{3\sqrt{10x^7y}}$

$= \dfrac{2\sqrt{5x^4y^2 \times 6y^6}}{3\sqrt{4x^3y^5 \times 10x^7y}}$

$= \dfrac{2}{3}\sqrt{\dfrac{30x^4y^8}{40x^{10}y^6}}$

$= \dfrac{2}{3}\sqrt{\dfrac{3y^2}{4x^6}}$

$= \dfrac{2}{3} \times \dfrac{\sqrt{3y^2}}{\sqrt{(2x^3)^2}}$

$= \dfrac{2}{3} \times \dfrac{y\sqrt{3}}{2x^3}$

$= \dfrac{y\sqrt{3}}{3x^3}$

# Rationalizing The Denominator

## Multiple Choice

**1** **Option D**

$$\frac{1}{\sqrt{2}} \times \frac{\sqrt{2}}{\sqrt{2}}$$

$$= \frac{\sqrt{2}}{2}$$

**2** **Option B**

$$\frac{4}{\sqrt{3}} \times \frac{\sqrt{3}}{\sqrt{3}}$$

$$= \frac{4\sqrt{3}}{3}$$

**3** **Option B**

$$\frac{3}{\sqrt{12}}$$

$$= \frac{3}{\sqrt{4 \times 3}}$$

$$= \frac{3}{2\sqrt{3}}$$

$$\therefore \sqrt{3}$$

**4** **Option A**

$$\frac{10}{\sqrt{18}}$$

$$= \frac{10}{\sqrt{9 \times 2}} = \frac{10}{3\sqrt{2}}$$

$$= \frac{10}{3\sqrt{2}} \times \frac{\sqrt{2}}{\sqrt{2}}$$

$$= \frac{10\sqrt{2}}{3 \times 2} = \frac{10\sqrt{2}}{6}$$

$$= \frac{5\sqrt{2}}{3}$$

**5** **Option C**

$$\frac{\sqrt{20}}{\sqrt{50}}$$

$$= \frac{\sqrt{4 \times 5}}{\sqrt{25 \times 2}} = \frac{2\sqrt{5}}{5\sqrt{2}}$$

$$= \frac{2\sqrt{5}}{5\sqrt{2}} \times \frac{\sqrt{2}}{\sqrt{2}}$$

$$= \frac{2\sqrt{10}}{5 \times 2}$$

$$= \frac{2\sqrt{10}}{10}$$

$$= \frac{\sqrt{10}}{5}$$

## Time To Practice

**1** **a** $\frac{1}{\sqrt{7}}$

$= \frac{1}{\sqrt{7}} \times \frac{\sqrt{7}}{\sqrt{7}}$

$= \frac{\sqrt{7}}{(\sqrt{7})^2}$

$= \frac{\sqrt{7}}{7}$

**b** $\frac{4}{\sqrt{5}}$

$= \frac{4}{\sqrt{5}} \times \frac{\sqrt{5}}{\sqrt{5}}$

$= \frac{4\sqrt{7}}{(\sqrt{5})^2}$

$= \frac{4\sqrt{5}}{5}$

**c** $\frac{8}{\sqrt{2}}$

$= \frac{8}{\sqrt{2}} \times \frac{\sqrt{2}}{\sqrt{2}}$

$= \frac{8\sqrt{2}}{(\sqrt{2})^2}$

$= \frac{8\sqrt{2}}{2}$

$= 4\sqrt{2}$

**d** $\frac{\sqrt{5}}{\sqrt{3}}$

$= \frac{\sqrt{5}}{\sqrt{3}} \times \frac{\sqrt{3}}{\sqrt{3}}$

$= \frac{\sqrt{5 \times 3}}{(\sqrt{3})^2}$

$= \frac{\sqrt{15}}{3}$

**e** $\frac{\sqrt{7}}{\sqrt{2}}$

$= \frac{\sqrt{7}}{\sqrt{2}} \times \frac{\sqrt{2}}{\sqrt{2}}$

$= \frac{\sqrt{7 \times 2}}{(\sqrt{2})^2}$

$= \frac{\sqrt{14}}{2}$

**f** $\frac{\sqrt{11}}{\sqrt{2}}$

$= \frac{\sqrt{11}}{\sqrt{2}} \times \frac{\sqrt{2}}{\sqrt{2}}$

$= \frac{\sqrt{11 \times 2}}{(\sqrt{2})^2}$

$= \frac{\sqrt{22}}{2}$

**2** **a** $\frac{4\sqrt{2}}{\sqrt{7}}$

$= \frac{4\sqrt{2}}{\sqrt{7}} \times \frac{\sqrt{7}}{\sqrt{7}}$

$= \frac{4\sqrt{14}}{(\sqrt{7})^2}$

$= \frac{4\sqrt{14}}{7}$

**b** $\frac{5\sqrt{2}}{\sqrt{3}}$

$= \frac{5\sqrt{2}}{\sqrt{3}} \times \frac{\sqrt{3}}{\sqrt{3}}$

$= \frac{5\sqrt{6}}{(\sqrt{3})^2}$

$= \frac{5\sqrt{6}}{3}$

**c** $\frac{3\sqrt{6}}{\sqrt{7}}$

$= \frac{3\sqrt{6}}{\sqrt{7}} \times \frac{\sqrt{7}}{\sqrt{7}}$

$= \frac{3\sqrt{42}}{(\sqrt{7})^2}$

$= \frac{3\sqrt{42}}{7}$

**d** $\frac{2\sqrt{7}}{5\sqrt{10}}$

$= \frac{2\sqrt{7}}{5\sqrt{10}} \times \frac{\sqrt{10}}{\sqrt{10}}$

$= \frac{2\sqrt{70}}{5(\sqrt{10})^2}$

$= \frac{2\sqrt{70}}{5 \times 10}$

$= \frac{\sqrt{70}}{25}$

**e** $\frac{4\sqrt{7}}{5\sqrt{3}}$

$= \frac{4\sqrt{7}}{5\sqrt{3}} \times \frac{\sqrt{3}}{\sqrt{3}}$

$= \frac{4\sqrt{21}}{5(\sqrt{3})^2}$

$= \frac{4\sqrt{21}}{5 \times 3}$

$= \frac{4\sqrt{21}}{15}$

**f** $\frac{4\sqrt{3}}{3\sqrt{5}}$

$= \frac{4\sqrt{3}}{3\sqrt{5}} \times \frac{\sqrt{5}}{\sqrt{5}}$

$= \frac{4\sqrt{15}}{3(\sqrt{5})^2}$

$= \frac{4\sqrt{15}}{3 \times 5}$

$= \frac{4\sqrt{15}}{15}$

**2** **g** $\frac{8\sqrt{3}}{7\sqrt{7}}$

$= \frac{8\sqrt{3}}{7\sqrt{7}} \times \frac{\sqrt{7}}{\sqrt{7}}$

$= \frac{8\sqrt{21}}{7(\sqrt{7})^2}$

$= \frac{8\sqrt{21}}{7 \times 7}$

$= \frac{8\sqrt{21}}{49}$

**h** $\frac{9\sqrt{6}}{2\sqrt{3}}$

$= \frac{9\sqrt{6}}{2\sqrt{3}} \times \frac{\sqrt{3}}{\sqrt{3}}$

$= \frac{9\sqrt{18}}{2 \times 3}$

$= \frac{9\sqrt{9 \times 2}}{6}$

$= \frac{9 \times 3\sqrt{2}}{6}$

$= \frac{9\sqrt{2}}{2}$

**i** $\frac{5\sqrt{12}}{9\sqrt{6}}$

$= \frac{5\sqrt{12}}{9\sqrt{6}} \times \frac{\sqrt{6}}{\sqrt{6}}$

$= \frac{5\sqrt{72}}{9 \times 6}$

$= \frac{5\sqrt{36 \times 2}}{54}$

$= \frac{5 \times 6\sqrt{2}}{54}$

$= \frac{5\sqrt{2}}{9}$

**3** **a** $\dfrac{4\sqrt{10}}{\sqrt{18}}$

$= \dfrac{4\sqrt{10}}{\sqrt{9 \times 2}} = \dfrac{4\sqrt{10}}{3\sqrt{2}}$

$= \dfrac{4\sqrt{10}}{3\sqrt{2}} \times \dfrac{\sqrt{2}}{\sqrt{2}}$

$= \dfrac{4\sqrt{20}}{6}$

$= \dfrac{2\sqrt{4 \times 5}}{3}$

$= \dfrac{4\sqrt{5}}{3}$

**b** $\dfrac{8\sqrt{20}}{\sqrt{28}}$

$= \dfrac{8\sqrt{4 \times 5}}{\sqrt{4 \times 7}}$

$= \dfrac{16\sqrt{5}}{2\sqrt{7}} = \dfrac{8\sqrt{5}}{\sqrt{7}}$

$= \dfrac{8\sqrt{5}}{\sqrt{7}} \times \dfrac{\sqrt{7}}{\sqrt{7}}$

$= \dfrac{8\sqrt{35}}{7}$

**c** $\dfrac{2\sqrt{45}}{5\sqrt{12}}$

$= \dfrac{2\sqrt{9 \times 5}}{5\sqrt{4 \times 3}} = \dfrac{6\sqrt{5}}{10\sqrt{3}} = \dfrac{3\sqrt{5}}{5\sqrt{3}}$

$= \dfrac{3\sqrt{5}}{5\sqrt{3}} \times \dfrac{\sqrt{3}}{\sqrt{3}}$

$= \dfrac{3\sqrt{15}}{15}$

$= \dfrac{\sqrt{15}}{5}$

**d** $\dfrac{5\sqrt{50}}{3\sqrt{75}}$

$= \dfrac{5\sqrt{25 \times 2}}{3\sqrt{25 \times 3}}$

$= \dfrac{25\sqrt{2}}{15\sqrt{3}} = \dfrac{5\sqrt{2}}{3\sqrt{3}}$

$= \dfrac{5\sqrt{2}}{3\sqrt{3}} \times \dfrac{\sqrt{3}}{\sqrt{3}}$

$= \dfrac{5\sqrt{6}}{9}$

**e** $\dfrac{6\sqrt{44}}{2\sqrt{20}}$

$= \dfrac{3\sqrt{4 \times 11}}{\sqrt{4 \times 5}} = \dfrac{6\sqrt{11}}{2\sqrt{5}} = \dfrac{3\sqrt{11}}{\sqrt{5}}$

$= \dfrac{3\sqrt{11}}{\sqrt{5}} \times \dfrac{\sqrt{5}}{\sqrt{5}}$

$= \dfrac{3\sqrt{55}}{5}$

**f** $\dfrac{5\sqrt{63}}{8\sqrt{32}}$

$= \dfrac{5\sqrt{9 \times 7}}{8\sqrt{16 \times 2}} = \dfrac{15\sqrt{7}}{32\sqrt{2}}$

$= \dfrac{15\sqrt{7}}{32\sqrt{2}} \times \dfrac{\sqrt{2}}{\sqrt{2}}$

$= \dfrac{15\sqrt{14}}{64}$

**3** **g** $\dfrac{3\sqrt{128}}{4\sqrt{45}}$

$= \dfrac{3\sqrt{64 \times 2}}{4\sqrt{9 \times 5}}$

$= \dfrac{24\sqrt{2}}{12\sqrt{5}} = \dfrac{2\sqrt{2}}{\sqrt{5}}$

$= \dfrac{2\sqrt{2}}{\sqrt{5}} \times \dfrac{\sqrt{5}}{\sqrt{5}}$

$= \dfrac{2\sqrt{10}}{5}$

**h** $\dfrac{3\sqrt{125}}{5\sqrt{48}}$

$= \dfrac{3\sqrt{25 \times 5}}{5\sqrt{16 \times 3}}$

$= \dfrac{15\sqrt{5}}{20\sqrt{3}} = \dfrac{3\sqrt{5}}{4\sqrt{3}}$

$= \dfrac{3\sqrt{5}}{4\sqrt{3}} \times \dfrac{\sqrt{3}}{\sqrt{3}}$

$= \dfrac{3\sqrt{15}}{12}$

$= \dfrac{\sqrt{15}}{4}$

**i** $\dfrac{2\sqrt{98}}{7\sqrt{27}}$

$= \dfrac{2\sqrt{49 \times 2}}{7\sqrt{9 \times 3}} = \dfrac{14\sqrt{2}}{21\sqrt{3}} = \dfrac{2\sqrt{2}}{3\sqrt{3}}$

$= \dfrac{2\sqrt{2}}{3\sqrt{3}} \times \dfrac{\sqrt{3}}{\sqrt{3}}$

$= \dfrac{2\sqrt{6}}{9}$

$= \dfrac{2\sqrt{6}}{9}$

## Challenge Yourself

**1** **a** $\frac{3\sqrt{3}}{2} \times \frac{5}{\sqrt{6}}$

$= \frac{15\sqrt{3}}{2\sqrt{6}}$

$= \frac{15\sqrt{3}}{2\sqrt{6}} \times \frac{\sqrt{6}}{\sqrt{6}} = \frac{15\sqrt{18}}{12}$

$= \frac{5\sqrt{9 \times 2}}{4}$

$= \frac{15\sqrt{2}}{4}$

**b** $\frac{5\sqrt{18}}{3} \times \frac{\sqrt{5}}{6\sqrt{3}}$

$= \frac{5\sqrt{9 \times 2}}{3} \times \frac{\sqrt{5}}{6\sqrt{3}}$

$= \frac{15\sqrt{2}}{3} \times \frac{\sqrt{5}}{6\sqrt{3}}$

$= \frac{15\sqrt{10}}{18\sqrt{3}} = \frac{5\sqrt{10}}{6\sqrt{3}}$

$= \frac{5\sqrt{10}}{6\sqrt{3}} \times \frac{\sqrt{3}}{\sqrt{3}}$

$= \frac{5\sqrt{30}}{18}$

**c** $\frac{\sqrt{20}}{5\sqrt{2}} \times \frac{2}{\sqrt{27}}$

$= \frac{\sqrt{4 \times 5}}{5\sqrt{2}} \times \frac{2}{\sqrt{9 \times 3}}$

$= \frac{2\sqrt{5}}{5\sqrt{2}} \times \frac{2}{3\sqrt{3}} = \frac{4\sqrt{5}}{15\sqrt{6}}$

$= \frac{4\sqrt{5}}{15\sqrt{6}} \times \frac{\sqrt{6}}{\sqrt{6}}$

$= \frac{4\sqrt{30}}{15 \times 6}$

$= \frac{2\sqrt{30}}{15 \times 3}$

$= \frac{2\sqrt{30}}{45}$

**d** $\frac{\sqrt{12}}{3\sqrt{5}} \times \frac{4\sqrt{15}}{\sqrt{2}}$

$= \frac{\sqrt{4 \times 3}}{3\sqrt{5}} \times \frac{4\sqrt{15}}{\sqrt{2}}$

$= \frac{2\sqrt{3}}{3\sqrt{5}} \times \frac{4\sqrt{15}}{\sqrt{2}}$

$= \frac{8\sqrt{45}}{3\sqrt{10}} = \frac{8\sqrt{9 \times 5}}{3\sqrt{10}}$

$= \frac{24\sqrt{5}}{3\sqrt{10}} = \frac{8\sqrt{5}}{\sqrt{10}}$

$= \frac{8\sqrt{5}}{\sqrt{10}} \times \frac{\sqrt{10}}{\sqrt{10}}$

$= \frac{8\sqrt{50}}{10}$

$= \frac{8\sqrt{25 \times 2}}{10} = \frac{40\sqrt{2}}{10}$

$= 4\sqrt{2}$

**2** **a** $\dfrac{3\sqrt{28}}{\sqrt{18}} \div \dfrac{9\sqrt{12}}{2\sqrt{4}}$

$= \dfrac{3\sqrt{28}}{\sqrt{18}} \times \dfrac{2\sqrt{4}}{9\sqrt{12}}$

$= \dfrac{3\sqrt{4 \times 7}}{\sqrt{9 \times 2}} \times \dfrac{4}{9\sqrt{4 \times 3}}$

$= \dfrac{6\sqrt{7}}{3\sqrt{2}} \times \dfrac{4}{18\sqrt{3}}$

$= \dfrac{\sqrt{7}}{\sqrt{2}} \times \dfrac{4}{9\sqrt{3}} = \dfrac{4\sqrt{7}}{9\sqrt{6}}$

$= \dfrac{4\sqrt{7}}{9\sqrt{6}} \times \dfrac{\sqrt{6}}{\sqrt{6}}$

$= \dfrac{4\sqrt{42}}{54}$

$= \dfrac{2\sqrt{42}}{27}$

**b** $\dfrac{4\sqrt{3}}{\sqrt{72}} \div \dfrac{3\sqrt{5}}{2\sqrt{12}}$

$= \dfrac{4\sqrt{3}}{\sqrt{72}} \times \dfrac{2\sqrt{12}}{3\sqrt{5}}$

$= \dfrac{4\sqrt{3}}{\sqrt{36 \times 2}} \times \dfrac{2\sqrt{4 \times 3}}{3\sqrt{5}}$

$= \dfrac{4\sqrt{3}}{6\sqrt{2}} \times \dfrac{4\sqrt{3}}{3\sqrt{5}}$

$= \dfrac{4 \times 4 \times 3}{18\sqrt{10}} = \dfrac{48}{18\sqrt{10}}$

$= \dfrac{8}{3\sqrt{10}} \times \dfrac{\sqrt{10}}{\sqrt{10}}$

$= \dfrac{8\sqrt{10}}{30}$

$= \dfrac{4\sqrt{10}}{15}$

**3** **a** $\dfrac{\sqrt{125} - \sqrt{80}}{\sqrt{98}}$

$= \dfrac{\sqrt{25 \times 5} - \sqrt{16 \times 5}}{\sqrt{49 \times 2}}$

$= \dfrac{5\sqrt{5} - 4\sqrt{5}}{7\sqrt{2}}$

$= \dfrac{\sqrt{5}}{7\sqrt{2}}$

$= \dfrac{\sqrt{5}}{7\sqrt{2}} \times \dfrac{\sqrt{2}}{\sqrt{2}}$

$= \dfrac{\sqrt{10}}{14}$

**b** $\dfrac{\sqrt{75} - \sqrt{27}}{\sqrt{90}}$

$= \dfrac{\sqrt{25 \times 3} - \sqrt{9 \times 3}}{\sqrt{9 \times 10}}$

$= \dfrac{5\sqrt{3} - 3\sqrt{3}}{3\sqrt{10}} = \dfrac{2\sqrt{3}}{3\sqrt{10}}$

$= \dfrac{2\sqrt{3}}{3\sqrt{10}} \times \dfrac{\sqrt{10}}{\sqrt{10}}$

$= \dfrac{2\sqrt{30}}{30}$

$= \dfrac{\sqrt{30}}{15}$

Linear
Equations

# Basic Equations

## Multiple Choice

**1** **Option D**

$$7x = 28$$

$$x = \frac{28}{7}$$

$$= 4$$

**2** **Option C**

$$\frac{5x}{3} = 10$$

$$5x = 10 \times 3 = 30$$

$$x = \frac{30}{5}$$

$$x = 6$$

**3** **Option B**

$$3x + 5 = 11$$

$$3x = 11 - 5 = 6$$

$$x = \frac{6}{3}$$

$$x = 2$$

**4** **Option D**

$$\frac{x}{6} - 1 = 2$$

$$\frac{x}{6} = 2 + 1 = 3$$

$$x = 3 \times 6$$

$$x = 18$$

**5** **Option A**

$$\frac{3x}{4} + \frac{1}{2} = \frac{7}{2}$$

$$\frac{3x}{4} = \frac{7}{2} - \frac{1}{2} = \frac{6}{2} = 3$$

$$3x = 3 \times 4 = 12$$

$$x = \frac{12}{3}$$

$$x = 4$$

## Time To Practice

**1** **a**
$x + 6 = 8$
$x = 8 - 6$
$x = 2$

**b**
$x - 7 = 4$
$x = 4 + 7$
$x = 11$

**c**
$8 - x = 3$
$-x = 3 - 8$
$-x = -5$
$x = 5$

**d**
$6a = -42$
$a = \frac{-42}{6}$
$a = -7$

**e**
$9a = 63$
$a = \frac{63}{9}$
$a = 7$

**f**
$4a = -44$
$a = \frac{-44}{4}$
$a = -11$

**g**
$\frac{m}{4} = 3$
$m = 3 \times 4$
$m = 12$

**h**
$\frac{3m}{5} = 12$
$3m = 12 \times 5$
$3m = 60$
$m = \frac{60}{3}$
$m = 20$

**i**
$\frac{7m}{10} = -7$
$7m = -7 \times 10$
$7m = -70$
$m = \frac{-70}{7}$
$m = -10$

**2** **a**
$2x + 9 = 15$
$2x = 15 - 9 = 6$
$2x = 6$
$x = \frac{6}{2}$
$x = 3$

**b**
$4x + 3 = 23$
$4x = 23 - 3 = 20$
$4x = 20$
$x = \frac{20}{4}$
$x = 5$

**c**
$3x + 5 = -4$
$3x = -4 - 5$
$3x = -9$
$x = \frac{-9}{3}$
$x = -3$

**2** **d** $6b - 12 = 48$

$6b = 48 + 12$
$6b = 60$

$b = \frac{60}{6}$

$b = 10$

**e** $-3b + 5 = 17$

$-3b = 17 - 5$
$-3b = 12$

$b = \frac{12}{-3}$

$b = -4$

**f** $-2b + 6 = 4$

$-2b = 4 - 6$
$-2b = -2$

$b = \frac{-9}{-2}$

$b = 1$

**g** $-4k - 9 = 3$

$-4k = 3 + 9$
$-4k = 12$

$-4k = \frac{12}{-4}$

$k = -3$

**h** $-3k - 6 = -3$

$-3k = -3 + 6$
$-3k = 3$

$k = \frac{3}{-3}$

$k = -1$

**i** $6 - 5k = 16$

$-5k = 16 - 6$
$-5k = 10$

$k = \frac{10}{-5}$

$x = -2$

**3** **a** $\frac{x}{2} = 10$

$x = 10 \times 2$

$x = 20$

**b** $\frac{4x}{7} = -4$

$4x = -4 \times 7$

$4x = -28$

$x = \frac{-28}{4}$

$x = -7$

**c** $-\frac{2x}{3} = -6$

$-2x = -6 \times 3$

$-2x = -18$

$x = \frac{-18}{-2}$

$x = 9$

**d** $\frac{y}{2} - 5 = 3$

$\frac{y}{2} = 3 + 5$

$\frac{y}{2} = 8$

$y = 8 \times 2$
$y = 16$

**e** $\frac{3y}{2} + 2 = 8$

$\frac{3y}{2} = 8 - 2 = 6$

$3y = 6 \times 2 = 12$

$y = \frac{12}{3}$

$y = 4$

**f** $\frac{3y}{5} - 1 = 8$

$\frac{3y}{5} = 8 + 1 = 9$

$3y = 9 \times 5 = 45$

$y = \frac{45}{3}$

$y = 15$

**3** **g** $-\frac{4g}{7} - 1 = 11$

$-\frac{4g}{7} = 11 + 1 = 12$

$-4g = 12 \times 7 = 84$

$g = \frac{84}{-4}$

$g = -21$

**h** $6 - \frac{4x}{5} = -2$

$-\frac{4x}{5} = -2 - 6 = -8$

$-4x = -8 \times 5 = -40$

$x = \frac{-40}{-4}$

$x = 10$

**i** $10 - \frac{7x}{9} = 3$

$-\frac{7x}{9} = 3 - 10 = -7$

$-7x = -7 \times 9 = -63$

$x = \frac{-63}{-7}$

$g = 9$

**j** $\frac{3x}{2} + 2 = 8$

$\frac{3x}{2} = 8 - 2 = 6$

$3x = 6 \times 2 = 12$

$x = \frac{12}{3}$

$x = 4$

**k** $5 - \frac{y}{2} = 1$

$-\frac{y}{2} = 1 - 5 = -4$

$-y = -4 \times 2 = -8$

$y = \frac{-8}{-1}$

$y = 8$

**l** $4 - \frac{2z}{3} = 0$

$\frac{2z}{3} = 0 - 4 = -4$

$2z = -4 \times 3 = -12$

$z = \frac{-12}{2}$

$x = 6$

## Challenge Yourself

**1** **a**

$$2a - \frac{3}{5} = 1$$

$$2a = 1 + \frac{3}{5}$$

$$2a = \frac{5}{5} + \frac{3}{5} = \frac{8}{5}$$

$$a = \frac{8}{5} \div 2$$

$$a = \frac{8}{5} \times \frac{1}{2}$$

$$a = \frac{8}{10}$$

$$a = \frac{4}{5}$$

**b**

$$\frac{a}{3} - 2 = \frac{1}{4}$$

$$\frac{a}{3} = \frac{1}{4} + 2$$

$$\frac{a}{3} = \frac{1}{4} + \frac{8}{4} = \frac{9}{4}$$

$$\frac{a}{3} = \frac{9}{4} \times 3$$

$$a = \frac{27}{4}$$

**c**

$$\frac{5k}{6} + 4 = -\frac{1}{3}$$

$$\frac{5k}{6} = -\frac{1}{3} - 4 = -\frac{1}{3} - \frac{12}{3} = -\frac{13}{3}$$

$$5k = -\frac{13}{3} \times 6$$

$$5k = -26$$

$$k = -\frac{26}{5}$$

**d**

$$\frac{7k}{9} - 1 = \frac{2}{3}$$

$$\frac{7k}{9} = \frac{2}{3} + 1 = \frac{2}{3} + \frac{3}{3} = \frac{5}{3}$$

$$\frac{7k}{9} = \frac{5}{3}$$

$$7k = \frac{5}{3} \times 9 = 15$$

$$k = \frac{15}{7}$$

1 e

$$-\frac{3p}{8}+\frac{1}{2}=\frac{5}{6}$$

$$-\frac{3p}{8}=\frac{5}{6}-\frac{1}{2}$$

$$-\frac{3p}{8}=\frac{5}{6}-\frac{3}{6}=\frac{2}{6}=\frac{1}{3}$$

$$-3p=\frac{1}{3}\times 8=\frac{8}{3}$$

$$p=\frac{8}{3}\div -3$$

$$p=\frac{8}{3}\times\frac{1}{-3}$$

$$p=-\frac{8}{9}$$

f

$$-\frac{2p}{3}-\frac{1}{9}=-\frac{3}{2}$$

$$-\frac{2p}{3}=-\frac{3}{2}+\frac{1}{9}$$

$$-\frac{2p}{3}=\frac{-27+2}{18}=-\frac{25}{18}$$

$$-2p=-\frac{25}{18}\times 3=-\frac{25}{6}$$

$$p=-\frac{25}{6}\div -2$$

$$p=-\frac{25}{6}\times\frac{1}{-2}$$

$$p=\frac{25}{12}$$

2

$$3x-p=q$$

$$3x=q+p$$

$$x=\frac{q+p}{3}$$

3

$$y=mx+c$$

$$mx=y-c$$

$$x=\frac{y-c}{m}$$

4

$$ab-c=d$$

$$ab=d+c$$

$$a=\frac{d+c}{b}$$

# Variables on Both Sides

## Multiple Choice

**1** **Option A**

$3x + 5 = 2x + 12$

$3x - 2x = 12 - 5$

$x = 7$

**2** **Option B**

$9y - 7 = 6y + 11$

$9y - 6y = 11 + 7$

$3y = 18$

$y = 6$

**3** **Option B**

$7z - 3 = 5z + 13$

$7z - 5z = 13 + 3$

$2z = 16$

$z = 8$

**4** **Option C**

$4a + 8 = 2a + 18$

$4a - 2a = 18 - 8$

$2a = 10$

$a = 5$

**5** **Option D**

$8b + 3 = 4b + 15$

$8b - 4b = 15 - 3$

$4b = 12$

$b = 3$

## Time To Practice

**1** **a**
$6 - x = 5 - 2x$
$-x + 2x = 5 - 6$
$x = -1$

**b**
$8 - 2x = 10 - 3x$
$-2x + 3x = 10 - 8$
$x = 2$

**c**
$8x + 45 = 6x + 15$
$8x - 6x = 15 - 45$
$2x = -30$
$x = -15$

**d**
$7a + 20 = 10a - 10$
$20 + 10 = 10a - 7a$
$30 = 3a$
$a = 10$

**e**
$3 - g = 17 - 3g$
$-g + 3g = 17 - 3$
$2g = 14$
$g = 7$

**f**
$w + 8 = 14 - 5w$
$w + 5w = 14 - 8$
$6w = 6$
$w = 1$

**g**
$15 - p = 27 - 3p$
$-p + 3p = 27 - 15$
$2p = 12$
$p = 6$

**h**
$14g = 2g + 60$
$14g - 2g = 60$
$12g = 60$
$g = 5$

**i**
$3h - 1 = 23 - h$
$3h + h = 23 + 1$
$4h = 24$
$h = 6$

**2** **a**
$6w + 27 = 11w + 17$
$27 - 17 = 11w - 6w$
$10 = 5w$
$w = 2$

**b**
$5t + 4 = 10t - 11$
$4 + 11 = 10t - 5t$
$15 = 5t$
$t = 3$

**2** **c** $12r - 20 = 15r - 38$

$-20 + 38 = 15r - 12r$

$18 = 3r$

$r = 6$

**d** $20f - 4 = 7f + 35$

$20f - 7f = 35 + 4$

$13f = 39$

$f = 3$

**e** $-4q - 30 = 10q + 110$

$-30 - 110 -= 10q + 4q$

$-140 = 14q$

$q = -10$

**f** $90 - 10v = 2 - 21v$

$-10v + 21v = 2 - 90$

$11v = -88$

$v = -8$

**g** $17 - 3n = 52 - 10n$

$-3n + 10n = 52 - 17$

$7n = 35$

$n = 5$

**h** $-5m - 9 = 27 - 11m$

$-5m + 11m = 27 + 9$

$6m = 36$

$m = 6$

**i** $2 - 6x = 10 - 4x$

$2 - 10 = -4x + 6x$

$-8 = 2x$

$x = -4$

**j** $-11t - 4 = 60 - 3t$

$-4 - 60 = -3t + 11t$

$-64 = 8t$

$t = -8$

## Challenge Yourself

**1** **a** $3x + 4 = 2x + \frac{5}{2}$

$3x - 2x = \frac{5}{2} - 4$

$x = \frac{5}{2} - \frac{8}{2}$

$x = -\frac{3}{2}$

**b** $5x - \frac{3}{4} = 2x + \frac{1}{2}$

$5x - 2x = \frac{1}{2} + \frac{3}{4}$

$3x = \frac{2}{4} + \frac{3}{4} = \frac{5}{4}$

$x = \frac{5}{4} \times \frac{1}{3}$

$x = \frac{5}{12}$

**1** **c**

$$\frac{7q}{3}+1=\frac{5q}{3}+\frac{4}{5}$$

$$\frac{7q}{3}-\frac{5q}{3}=\frac{4}{5}-1$$

$$\frac{2q}{3}=\frac{4}{5}-\frac{5}{5}=-\frac{1}{5}$$

$$2q=-\frac{1}{5}\times 3=-\frac{3}{5}$$

$$q=-\frac{3}{5}\times\frac{1}{2}$$

$$q=-\frac{3}{10}$$

**d**

$$\frac{8p}{5}-\frac{2}{3}=\frac{4p}{5}-\frac{1}{6}$$

$$\frac{8p}{5}-\frac{4p}{5}=-\frac{1}{6}+\frac{2}{3}$$

$$\frac{4p}{5}=-\frac{1}{6}+\frac{4}{6}=\frac{3}{6}=\frac{1}{2}$$

$$4p=\frac{1}{2}\times 5=\frac{5}{2}$$

$$p=\frac{5}{2}\times\frac{1}{4}$$

$$p=\frac{5}{8}$$

**e**

$$-\frac{s}{3}-\frac{1}{5}=-\frac{2s}{3}+\frac{3}{10}$$

$$-\frac{s}{3}+\frac{2s}{3}=\frac{3}{10}+\frac{1}{5}$$

$$\frac{s}{3}=\frac{3}{10}+\frac{2}{10}=\frac{5}{10}=\frac{1}{2}$$

$$s=\frac{1}{2}\times 3$$

$$s=\frac{3}{2}$$

**f**

$$\frac{7t}{4}-\frac{1}{2}=\frac{5t}{4}+\frac{1}{3}$$

$$\frac{7t}{4}-\frac{5t}{4}=\frac{1}{3}+\frac{1}{2}$$

$$\frac{2t}{4}=\frac{2+3}{6}=\frac{5}{6}$$

$$\frac{t}{2}=\frac{5}{6}$$

$$t=\frac{5}{6}\times 2$$

$$t=\frac{5}{3}$$

# Equations With Brackets

## Multiple Choice

**1** **Option B**

$2(a + 3) = 10$

$a + 3 = 5$

$a = 2$

**2** **Option C**

$4(3x - 2) = 28$

$3x - 2 = 7$

$3x = 9$

$x = 3$

**3** **Option C**

$2(y + 5) = 3y - 1$

$2y + 10 = 3y - 1$

$10 + 1 = 3y - 2y$

$y = 11$

**4** **Option D**

$2(z - 3) + 3(z + 2) = 5$

$2z - 6 + 3z + 6 = 5$

$5z = 5$

$z = 1$

**5** **Option A**

$5(3b + 2) = 2(7b + 3)$

$15b + 10 = 14b + 6$

$15b - 14b = 6 - 10$

$b = -4$

## Time To Practice

**1**

**a**

$4(x+3)=16$

$x+3=4$

$x=1$

**b**

$2(x-3)=12$

$x-3=6$

$x=9$

**c**

$2(x-4)=18$

$x-4=9$

$x=13$

**d**

$3(1-2p)=27$

$1-2p=9$

$-2p=8$

$p=-4$

**e**

$4(2q-5)=-12$

$2q-5=-3$

$2q=2$

$q=1$

**f**

$2(5+4r)=-30$

$5+4r=-15$

$4r=-20$

$4=-5$

**g**

$3(m+5)=2m$

$3m+15=2m$

$3m-2m=-15$

$m=-15$

**h**

$8(2-n)=-6n$

$16-8n=-6n$

$16=-6n+8n$

$16=2n$

$n=8$

**i**

$5(7-2s)=-15s$

$35-10s=-15s$

$-10s+15s=-35$

$5s=-35$

$s=-7$

**2** **a** $2(k-3)=k-5$

$2k-6=k-5$

$2k-k=-5+6$

$k=1$

**b** $7(2-x)=8-x$

$14-7x=8-x$

$14-8=-x+7x$

$6=6x$

$x=1$

**c** $10-3c=2(4-c)$

$10-3c=8-2c$

$10-8=-2c+3c$

$c=2$

**d** $4r+7=3(2r-5)$

$4r+7=6r-15$

$7+15=6r-4r$

$22=2r$

$r=11$

**e** $2(3x-2)-3(x+1)=5$

$6x-4-3x-3=5$

$3x-7=5$

$3x=12$

$x=4$

**f** $3(2x+3)+2(x+4)=25$

$6x+9+2x+8=25$

$8x+17=25$

$8x=8$

$x=1$

**g** $2(2y-3)+3(4y-1)=23$

$4y-6+12y-3=23$

$16y-9=23$

$16y=32$

$y=2$

**h** $2(9-w)-3(w+16)=0$

$18-2w-3w-48=0$

$-30-5w=0$

$-5w=30$

$w=-6$

**i** $7(j+1)-2(2j-3)=10$

$7j+7-4j+6=10$

$3j+13=10$

$3j=-3$

$j=-1$

**j** $7(2s-1)-4(s+3)=-49$

$14s-7-4s-12=-49$

$10s-19=-49$

$10s=-30$

$s=-3$

**3** **a**

$5(u+3) = 3(u+9)$

$5u + 15 = 3u + 27$

$5u - 3u = 27 - 15$

$2u = 12$

$u = 6$

**b**

$8(v-1) = 4(v+3)$

$8v - 8 = 4v + 12$

$8v - 4v = 12 + 8$

$4v = 20$

$v = 5$

**c**

$9(2x-5) = 3(4x+7)$

$18x - 45 = 12x + 21$

$18x - 12x = 21 + 45$

$6x = 66$

$x = 11$

**d**

$5(2y+9) = 3(3y+13)$

$10y + 45 = 9y + 39$

$10y - 9y = 39 - 45$

$y = -6$

**e**

$5(l-3) = 3(2l-1)$

$5l - 15 = 6l - 3$

$-15 + 3 = 6l - 5l$

$l = -12$

**f**

$2(4k+1) = 5(3-k)$

$8k + 2 = 15 - 5k$

$8k + 5k = 15 - 2$

$13k = 13$

$k = 1$

**g**

$4(2d+5) = 3(d+15)$

$8d + 20 = 3d + 45$

$8d - 3d = 45 - 20$

$5d = 25$

$d = 5$

**h**

$3(4x-1) = 7(2x-7)$

$12x - 3 = 14x - 49$

$-3 + 49 = 14x - 12x$

$46 = 2x$

$x = 23$

**4** **a**

$4(2v-3) = 3(3v+2) - 5$

$8v - 12 = 9v + 6 - 5$

$8v - 12 = 9v + 1$

$-12 - 1 = 9v - 8v$

$v = -13$

**b**

$4(j+3) + 2 = 2(3j+5) - 8$

$4j + 12 + 2 = 6j + 10 - 8$

$4j + 14 = 6j + 2$

$14 - 2 = 6j - 4j$

$2j = 12$

$j = 6$

**4** **c** $5(2w+3)-4=3(4w+5)-2$

$10w+15-4=12w+15-2$

$10w+11=12w+1$

$11-13=12w-10w$

$-2=2w$

$w=-1$

**d** $2(x-4)+7x=3(x+2)+4x$

$2x-8+7x=3x+6+4x$

$9x-8=7x+6$

$9x-7x=6+8$

$2x=14$

$x=7$

**e** $3(2m+1)-m=2(m+3)+2m$

$6m+3-m=2m+6+2m$

$5m+3=4m+6$

$5m-4m=6-3$

$m=3$

**f** $8n-3(2n-4)=4(n+2)+2n$

$8n-6n+12=4n+8+2n$

$2n+12=6n+8$

$12-8=6n-2n$

$4n=4$

$n=1$

## Challenge Yourself

**1** **a** $3\left(\frac{3y}{5}-1\right)=15$

$\frac{3y}{5}-1=5$

$\frac{3y}{5}=6$

$3y=30$

$y=10$

**b** $11\left(8-\frac{z}{4}\right)=99$

$8-\frac{z}{4}=9$

$-\frac{z}{4}=9-8$

$-\frac{z}{4}=1$

$z=-4$

**1** **c** $\frac{2}{3}(3k-6)-3(k+1)=11+k$

$\frac{6k}{3}-\frac{12}{3}-3k-3=11+k$

$2k-4-3k-3=11+k$

$-k-7=11+k$

$-7-11=k+k$

$-18=2k$

$k=-9$

**d** $\frac{1}{2}(6z-8)+3(z+2)=4(2z-1)$

$\frac{6z}{2}-\frac{8}{2}+3z+6=8z-4$

$3z-4+3z+6=8z-4$

$6z+2=8z-4$

$2+4=8z-6z$

$6=2z$

$z=3$

**e** $4-2(x-6)=\frac{2x}{3}$

$4-2x+12=\frac{2x}{3}$

$16-2x=\frac{2x}{3}$

$16=\frac{2x}{3}+2x$

$16=\frac{2x}{3}+\frac{6x}{3}$

$16=\frac{8x}{3}$

$8x=16\times 3$

$8x=48$

$x=6$

**1** **f** $\frac{7x}{5} - \frac{4x}{10} = 2\left(x + \frac{9}{2}\right)$

$\frac{7x}{5} - \frac{2x}{5} = 2x + 9$

$\frac{5x}{5} = 2x + 9$

$x = 2x + 9$

$-9 = 2x - x$

$x = -9$

**g** $2(h + 5(h - 1)) = 110$

$(h + 5h - 5) = 55$

$6h - 5 = 55$

$6h = 60$

$h = 10$

**h** $-6(2t - 3(t - 4)) = 0$

$(2t - 3t + 12) = 0$

$(-t + 12) = 0$

$-t + 12 = 0$

$t = 12$

**i** $2(2(p - 3) + 3(1 - p)) = 0$

$(2p - 6 + 3 - 3p) = 0$

$-p - 3 = 0$

$-p = 3$

$p = -3$

**j** $6 - 3(2 - 3(y + 1)) = 18$

$6 - 3(2 - 3y - 3) = 18$

$6 - 3(-1 - 3y) = 18$

$6 + 3 + 9y = 18$

$9 + 9y = 18$

$9y = 9$

$y = 1$

# Equations With Fractions

## Multiple Choice

**1** **Option C**

$$\frac{x}{2}+\frac{3x}{2}=16$$

$$\frac{4x}{2}=16$$

$$2x=16$$

$$x=8$$

**2** **Option D**

$$\frac{4x}{3}+\frac{x}{3}=10$$

$$\frac{5x}{3}=10$$

$$5x=30$$

$$x=6$$

**3** **Option B**

$$\frac{2y+1}{3}=y-2$$

$$2y+1=3(y-2)$$

$$2y+1=3y-6$$

$$1+6=3y-2y$$

$$y=7$$

**4** **Option D**

$$\frac{2a+1}{3}=\frac{a+4}{2}$$

$$2(2a+1)=3(a+4)$$

$$4a+2=3a+12$$

$$4a-3a=12-2$$

$$a=10$$

**5** **Option C**

$$\frac{3b+4}{4}=\frac{3b}{2}-2$$

$$\frac{3b+4}{4}=\frac{3b}{2}-\frac{2}{1}$$

$$\frac{3b+4}{4}=\frac{3b-4}{2}$$

$$2(3b+4)=4(3b-4)$$

$$6b+8=12b-16$$

$$8+16=12b-6b$$

$$24=6b$$

$$b=4$$

## Time To Practice

**1** **a** $\dfrac{x+1}{2} = 3$

$x + 1 = 6$

$x = 5$

**b** $\dfrac{y-3}{3} = 4$

$y - 3 = 12$

$y = 15$

**c** $\dfrac{z-9}{3} = -1$

$z - 9 = -3$

$z = 6$

**d** $\dfrac{3x+5}{2} = 7$

$3x + 5 = 14$

$3x = 9$

$x = 3$

**e** $\dfrac{5y-12}{3} = 11$

$5y - 12 = 33$

$5y = 45$

$y = 9$

**f** $\dfrac{3z-4}{7} = -4$

$3z - 4 = -28$

$3z = -24$

$z = -8$

**g** $\dfrac{7x-12}{3} = -25$

$7x - 12 = -75$

$7x = -63$

$x = -9$

**h** $\dfrac{29-2y}{3} = 5$

$29 - 2y = 15$

$29 - 15 = 2y$

$14 = 2y$

$y = 7$

**1** **i** $\frac{100 - 5z}{3} = 30$

$100 - 5z = 90$

$100 - 90 = 5z$

$10 = 5z$

$z = 2$

**j** $\frac{8x + 4}{5} = 2x$

$8x + 4 = 10x$

$4 = 10x - 8x$

$4 = 2x$

$x = 2$

**k** $\frac{12 - 4y}{6} = -y$

$12 - 4y = -6y$

$-4y + 6y = -12$

$2y = -12$

$y = -6$

**l** $\frac{15 - z}{3} = -2z$

$15 - z = -6z$

$-z + 6z = -15$

$5z = -15$

$z = -3$

**2** **a** $\frac{3x - 1}{4} + 6 = 5$

$\frac{3x - 1}{4} + \frac{6}{1} = 5$

$\frac{(3x - 1) + 24}{4} = 5$

$\frac{3x + 23}{4} = 5$

$3x + 23 = 20$

$3x = -3$

$x = -1$

**b** $\frac{12 - 3y}{3} - 3 = 5$

$\frac{12 - 3y}{3} - \frac{3}{1} = 5$

$\frac{(12 - 3y) - 9}{3} = 5$

$\frac{3 - 3y}{3} = 5$

$3 - 3y = 15$

$-3y = 12$

$y = -4$

**2** **c** $\dfrac{35-2z}{7}-3=2$

$\dfrac{35-2z}{7}-\dfrac{3}{1}=2$

$\dfrac{(35-2z)-21}{7}=2$

$\dfrac{14-2z}{7}=2$

$14-2z=14$

$14-14=2z$

$0=2z$

$z=0$

**d** $\dfrac{7x+4}{3}-8=-9$

$\dfrac{7x+4}{3}-\dfrac{8}{1}=-9$

$\dfrac{(7x+4)-24}{3}=-9$

$\dfrac{7x-20}{3}=-9$

$7x-20=-27$

$7x=-7$

$x=-1$

**e** $\dfrac{3y+1}{4}-3=-5$

$\dfrac{3y+1}{4}-\dfrac{3}{1}=-5$

$\dfrac{(3y+1)-12}{4}=-5$

$\dfrac{3y-11}{4}=-5$

$3y-11=-20$

$3y=-9$

$y=-3$

**f** $\dfrac{4z-6}{3}-7=3$

$\dfrac{4z-6}{3}-\dfrac{7}{1}=3$

$\dfrac{(4z-6)-21}{3}=3$

$\dfrac{4z-27}{3}=3$

$4z-27=9$

$4z=36$

$z=9$

**3** **a**

$$\frac{x}{3} - \frac{x}{2} = 1$$
$$\frac{2x - 3x}{6} = 1$$
$$\frac{-x}{6} = 1$$
$$-x = 6$$
$$x = -6$$

**b**

$$\frac{y}{4} - \frac{y}{5} = 1$$
$$\frac{5y - 4y}{20} = 1$$
$$\frac{y}{20} = 1$$
$$y = 20$$

**c**

$$\frac{5z}{4} - \frac{z}{2} = -\frac{3}{4}$$
$$\frac{5z}{4} - \frac{2z}{4} = -\frac{3}{4}$$
$$\frac{5z - 2z}{4} = -\frac{3}{4}$$
$$\frac{3z}{4} = -\frac{3}{4}$$
$$3z = -3$$
$$z = -1$$

**d**

$$\frac{x}{4} - \frac{x}{7} = \frac{9}{14}$$
$$\frac{7x - 4x}{28} = \frac{9}{14}$$
$$\frac{3x}{28} = \frac{9}{14}$$
$$3x = \frac{9}{14} \times 28$$
$$3x = 18$$
$$x = 6$$

**e**

$$y + \frac{y}{7} = 8$$
$$\frac{y}{1} + \frac{y}{7} = 8$$
$$\frac{7y + y}{7} = 8$$
$$\frac{8y}{7} = 8$$
$$8y = 56$$
$$y = 7$$

**f**

$$-z - \frac{z}{4} = \frac{5}{4}$$
$$\frac{-z}{1} - \frac{z}{4} = \frac{5}{4}$$
$$\frac{-4z - z}{4} = \frac{5}{4}$$
$$\frac{-5z}{4} = \frac{5}{4}$$
$$-5z = 5$$
$$z = -1$$

**3** **g**

$$2x + \frac{3x}{5} = -13$$

$$\frac{2x}{1} + \frac{3x}{5} = -13$$

$$\frac{10x + 3x}{5} = -13$$

$$\frac{13x}{5} = -13$$

$$13x = -65$$

$$x = -5$$

**h**

$$6y - \frac{3y}{2} = 27$$

$$\frac{6y}{1} - \frac{3y}{2} = 27$$

$$\frac{12y - 3y}{2} = 27$$

$$\frac{9y}{2} = 27$$

$$9y = 54$$

$$y = 6$$

**i**

$$-4z + \frac{5z}{2} = -15$$

$$\frac{-4z}{1} + \frac{5z}{2} = -15$$

$$\frac{-8z + 5z}{2} = -15$$

$$\frac{-3z}{2} = -15$$

$$-3z = -30$$

$$z = 10$$

**4** **a**

$$\frac{2x + 1}{3} = x - 2$$

$$2x + 1 = 3(x - 2)$$

$$2x + 1 = 3x - 6$$

$$1 + 6 = 3x - 2x$$

$$x = 7$$

**b**

$$\frac{5y - 3}{2} = 2y + 4$$

$$5y - 3 = 2(2y + 4)$$

$$5y - 3 = 4y + 8$$

$$5y - 4y = 8 + 3$$

$$y = 11$$

**4** **c** $\frac{z+17}{5} = z + 1$

$z + 17 = 5(z + 1)$

$z + 17 = 5z + 5$

$17 - 5 = 5z - z$

$12 = 4z$

$z = 3$

**d** $4x - 9 = \frac{2x+3}{3}$

$3(4x - 9) = 2x + 3$

$12x - 27 = 2x + 3$

$12x - 2x = 3 + 27$

$10x = 30$

$x = 3$

**e** $y + 3 = \frac{9-y}{2}$

$2(y + 3) = 9 - y$

$2y + 6 = 9 - y$

$2y + y = 9 - 6$

$3y = 3$

$y = 1$

**f** $-3z - 2 = \frac{8z-33}{5}$

$5(-3z - 2) = 8z - 33$

$-15z - 10 = 8z - 33$

$-10 + 33 = 8z + 15z$

$23 = 23z$

$z = 1$

**5** **a** $\frac{3x+7}{5} = \frac{2x}{3}$

$3(3x + 7) = 10x$

$9x + 21 = 10x$

$21 = 10x - 9x$

$x = 21$

**b** $\frac{3y-10}{7} = \frac{2y}{5}$

$5(3y - 10) = 14y$

$15y - 50 = 14y$

$15y - 14y = 50$

$y = 50$

**c** $\frac{z-8}{3} = \frac{z}{6}$

$6(z - 8) = 3z$

$6z - 48 = 3z$

$6z - 3z = 48$

$3z = 48$

$z = 16$

**d** $\frac{x-5}{3} = \frac{x-7}{4}$

$4(x - 5) = 3(x - 7)$

$4x - 20 = 3x - 21$

$4x - 3x = -21 + 20$

$x = -1$

**5**

**e** $\frac{y+2}{6} = \frac{y-5}{3}$

$3(y+2) = 6(y-5)$

$3y + 6 = 6y - 30$

$6 + 30 = 6y - 3y$

$36 = 3y$

$y = 12$

**f** $\frac{z+3}{4} = \frac{27-z}{6}$

$6(z+3) = 4(27-z)$

$6z + 18 = 108 - 4z$

$6z + 4z = 108 - 18$

$10z = 90$

$z = 9$

**g** $\frac{9x+3}{2} = \frac{13x+7}{3}$

$3(9x+3) = 2(13x+7)$

$27x + 9 = 26x + 14$

$27x - 26x = 14 - 9$

$x = 5$

**h** $\frac{4y+9}{15} = \frac{y+2}{5}$

$5(4y+9) = 15(y+2)$

$20y + 45 = 15y + 30$

$20y - 15y = 30 - 45$

$5y = -15$

$y = -3$

**i** $\frac{2z+7}{5} = \frac{3-z}{4}$

$4(2z+7) = 5(3-z)$

$8z + 28 = 15 - 5z$

$8z + 5z = 15 - 28$

$13z = -13$

$z = -1$

**j** $\frac{x+11}{2} = \frac{x+14}{3}$

$3(x+11) = 2(x+14)$

$3x + 33 = 2x + 28$

$3x - 2x = 28 - 33$

$x = -5$

**k** $\frac{3-z}{2} = \frac{6(9-z)}{4}$

$4(3-z) = 12(9-z)$

$12 - 4z = 108 - 12z$

$-4z + 12z = 108 - 12$

$8z = 96$

$z = 12$

**l** $\frac{3y-4}{2} = \frac{102-5y}{4}$

$4(3y-4) = 2(102-5y)$

$12y - 16 = 204 - 10y$

$12y + 10y = 204 + 16$

$22y = 220$

$y = 10$

**6** **a**

$$\frac{x+3}{2} - 3 = \frac{x+4}{4}$$

$$\frac{x+3}{2} - \frac{3}{1} = \frac{x+4}{4}$$

$$\frac{(x+3)-6}{2} = \frac{x+4}{4}$$

$$\frac{x-3}{2} = \frac{x+4}{4}$$

$$4(x-3) = 2(x+4)$$

$$4x - 12 = 2x + 8$$

$$4x - 2x = 8 + 12$$

$$2x = 20$$

$$x = 10$$

**b**

$$\frac{x+3}{10} = 2 - \frac{5-x}{2}$$

$$\frac{x+3}{10} = \frac{2}{1} - \frac{5-x}{2}$$

$$\frac{x+3}{10} = \frac{4-(5-x)}{2}$$

$$\frac{x+3}{10} = \frac{x-1}{2}$$

$$2(x+3) = 10(x-1)$$

$$2x + 6 = 10x - 10$$

$$6 + 10 = 10x - 2x$$

$$16 = 8x$$

$$x = 2$$

**c**

$$\frac{x-1}{4} = \frac{x+3}{2} - 4$$

$$\frac{x-1}{4} = \frac{x+3}{2} - \frac{4}{1}$$

$$\frac{x-1}{4} = \frac{(x+3)-8}{2}$$

$$\frac{x-1}{4} = \frac{x-5}{2}$$

$$2(x-1) = 4(x-5)$$

$$2x - 2 = 4x - 20$$

$$-2 + 20 = 4x - 2x$$

$$18 = 2x$$

$$x = 9$$

**d**

$$\frac{4x+9}{15} = 1 + \frac{x-3}{5}$$

$$\frac{4x+9}{15} = \frac{5}{5} + \frac{x-3}{5}$$

$$\frac{4x+9}{15} = \frac{x+2}{5}$$

$$5(4x+9) = 15(x+2)$$

$$20x + 45 = 15x + 30$$

$$20x - 15x = 30 - 45$$

$$5x = -15$$

$$x = -3$$

**6** **e** $\frac{6x+7}{5}=\frac{5x+1}{4}+1$

$\frac{6x+7}{5}=\frac{5x+1}{4}+\frac{4}{4}$

$\frac{6x+7}{5}=\frac{5x+5}{4}$

$4(6x+7)=5(5x+5)$

$24x+28=25x+25$

$28-25=25x-24x$

$x=3$

**f** $\frac{2x+10}{3}=\frac{6-5x}{4}-2$

$\frac{2x+10}{3}=\frac{6-5x}{4}-\frac{2}{1}$

$\frac{2x+10}{3}=\frac{(6-5x)-8}{4}$

$\frac{2x+10}{3}=\frac{-2-5x}{4}$

$4(2x+10)=3(-2-5x)$

$8x+40=-6-15x$

$8x+15x=-6-40$

$23x=-46$

$x=-2$

**g** $\frac{3x-4}{5}=\frac{5x+9}{7}-3$

$\frac{3x-4}{5}=\frac{5x+9}{7}-\frac{3}{1}$

$\frac{3x-4}{5}=\frac{(5x+9)-21}{7}$

$\frac{3x-4}{5}=\frac{5x-12}{7}$

$7(3x-4)=5(5x-12)$

$21x-28=25x-60$

$-28+60=25x-21x$

$32=4x$

$x=8$

**h** $6-\frac{7-2x}{3}=\frac{1-2x}{9}$

$\frac{6}{1}-\frac{7-2x}{3}=\frac{1-2x}{9}$

$\frac{18-(7-2x)}{3}=\frac{1-2x}{9}$

$\frac{2x+11}{3}=\frac{1-2x}{9}$

$9(2x+11)=3(1-2x)$

$18x+99=3-6x$

$18x+6x=3-99$

$24x=-96$

$x=-4$

**7** **a** $\frac{5x-10}{3}-\frac{3x}{2}=\frac{2x}{3}$

$\frac{2(5x-10)-9x}{6}=\frac{2x}{3}$

$\frac{10x-20-9x}{6}=\frac{2x}{3}$

$\frac{x-20}{6}=\frac{2x}{3}$

$3(x-20)=12x$

$3x-60=12x$

$-60=12x-3x$

$-60=9x$

$x=\frac{-60}{9}=-\frac{20}{3}$

**b** $\frac{-3x+12}{6}+\frac{2x-6}{3}=-\frac{x}{2}$

$\frac{-3x+12}{6}+\frac{2(2x-6)}{6}=-\frac{x}{2}$

$\frac{-3x+12+4x-12}{6}=-\frac{x}{2}$

$\frac{x}{6}=-\frac{x}{2}$

$2x=-6x$

$2x+6x=0$

$8x=0$

$x=0$

**c** $\frac{2x-4}{3}+\frac{x+5}{6}=\frac{2-x}{4}$

$\frac{2(2x-4)}{6}+\frac{x+5}{6}=\frac{2-x}{4}$

$\frac{4x-8+x+5}{6}=\frac{2-x}{4}$

$\frac{5x-3}{6}=\frac{2-x}{4}$

$4(5x-3)=6(2-x)$

$20x-12=12-6x$

$20x+6x=12+12$

$26x=24$

$x=\frac{24}{26}=\frac{12}{13}$

**d** $\frac{-x-6}{5}+\frac{4x+7}{10}=\frac{3x+1}{5}$

$\frac{2(-x-6)}{10}+\frac{4x+7}{10}=\frac{2(3x+1)}{10}$

$\frac{-2x-12+4x+7}{10}=\frac{6x+2}{10}$

$\frac{2x-5}{10}=\frac{6x+2}{10}$

$2x-5=6x+2$

$-5-2=6x-2x$

$-7=4x$

$x=-\frac{7}{4}$

**7** **e**

$$\frac{5x+8}{5}-\frac{3x+6}{10}=\frac{x+1}{2}$$

$$\frac{2(5x+8)}{10}-\frac{3x+6}{10}=\frac{x+1}{2}$$

$$\frac{10x+16}{10}-\frac{3x+6}{10}=\frac{x+1}{2}$$

$$\frac{10x+16-(3x+6)}{10}=\frac{x+1}{2}$$

$$\frac{10x+16-3x-6}{10}=\frac{x+1}{2}$$

$$\frac{7x+10}{10}=\frac{x+1}{2}$$

$$2(7x+10)=10(x+1)$$

$$14x+20=10x+10$$

$$14x-10x=10-20$$

$$4x=-10$$

$$4x=-\frac{10}{4}=-\frac{5}{2}$$

**f**

$$\frac{3x-8}{3}+\frac{3-x}{6}=\frac{2x-5}{4}$$

$$\frac{2(3x-8)}{6}+\frac{3-x}{6}=\frac{2x-5}{4}$$

$$\frac{6x-16}{6}+\frac{3-x}{6}=\frac{2x-5}{4}$$

$$\frac{6x-16+3-x}{6}=\frac{2x-5}{4}$$

$$\frac{5x-13}{6}=\frac{2x-5}{4}$$

$$4(5x-13)=6(2x-5)$$

$$20x-52=12x-30$$

$$20x-12x=-30+52$$

$$8x=22$$

$$x=\frac{22}{8}=\frac{11}{4}$$

## Challenge Yourself

**1** **a**

$$\frac{7x}{5}-\frac{4x}{10}=2\left(x+\frac{9}{2}\right)$$

$$\frac{14x}{10}-\frac{4x}{10}=2x+9$$

$$\frac{10x}{10}=2x+9$$

$$x=2x+9$$

$$-9=2x-x$$

$$x=-9$$

**1** **b**

$$\frac{3}{4} - 2\left(a + \frac{1}{5}\right) = \frac{a}{3}$$

$$\frac{3}{4} - 2a - \frac{2}{5} = \frac{a}{3}$$

$$\frac{3}{4} - \frac{2}{5} = \frac{a}{3} + \frac{2a}{1}$$

$$\frac{15 - 8}{20} = \frac{a + 6a}{3}$$

$$\frac{7}{20} = \frac{7a}{3}$$

$$21 = 140a$$

$$a = \frac{21}{140} = \frac{7 \times 3}{7 \times 20}$$

$$a = \frac{3}{20}$$

**c**

$$\frac{41}{9} = \frac{5}{2}\left(y + \frac{2}{3}\right) - \frac{y}{3}$$

$$\frac{41}{9} = \frac{5y}{2} + \frac{5}{3} - \frac{y}{3}$$

$$\frac{41}{9} - \frac{5}{3} = \frac{5y}{2} - \frac{y}{3}$$

$$\frac{41}{9} - \frac{15}{9} = \frac{15y - 2y}{6}$$

$$\frac{26}{9} = \frac{13y}{6}$$

$$6 \times 26 = 9 \times 13y$$

$$\mathrm{y} = \frac{6 \times 26}{9 \times 13} = \frac{6 \times 2}{9 \times 1} = \frac{2 \times 2}{3 \times 1}$$

$$\mathrm{y} = \frac{4}{3}$$

**d**

$$-k - \frac{5}{4}\left(1 - \frac{8k}{3}\right) = -\frac{19}{4}$$

$$-k - \frac{5}{4}\left(1 - \frac{8k}{3}\right) = -\frac{19}{4}$$

$$-k - \frac{5}{4} + \frac{40k}{12} = -\frac{19}{4}$$

$$-k + \frac{10k}{3} = -\frac{19}{4} + \frac{5}{4}$$

$$\frac{-k}{1} + \frac{10k}{3} = -\frac{14}{4}$$

$$\frac{-3k}{3} + \frac{10k}{3} = -\frac{7}{2}$$

$$\frac{7k}{3} = -\frac{7}{2}$$

$$14k = -21$$

$$k = -\frac{21}{14} = -\frac{7 \times 3}{7 \times 2}$$

$$k = -\frac{3}{2}$$

1 e $\frac{1}{2}\left(1-\frac{2n}{3}\right)=\frac{n}{5}+\frac{1}{2}$

$\frac{1}{2}-\frac{n}{6}=\frac{2n+5}{10}$

$\frac{3}{6}-\frac{n}{6}=\frac{2n+5}{10}$

$\frac{3-n}{6}=\frac{2n+5}{10}$

$10(3-n)=6(2n+5)$

$30-10n=12n+30$

$30-30=12n+10n$

$0=22n$

$n=0$

f $\frac{7}{6}-\frac{4p}{3}=2\left(p+\frac{3}{2}\right)-\frac{3p}{2}$

$\frac{7}{6}-\frac{4p}{3}=2p+3-\frac{3p}{2}$

$\frac{7}{6}-\frac{8p}{6}=\frac{2p}{1}+\frac{3}{1}-\frac{3p}{2}$

$\frac{7-8p}{6}=\frac{4p}{2}+\frac{6}{2}-\frac{3p}{2}$

$\frac{7-8p}{6}=\frac{p+6}{2}$

$2(7-8p)=6(p+6)$

$14-16p=6p+36$

$14-36=6p+16p$

$-22=22p$

$p=-\frac{22}{22}$

$p=-1$

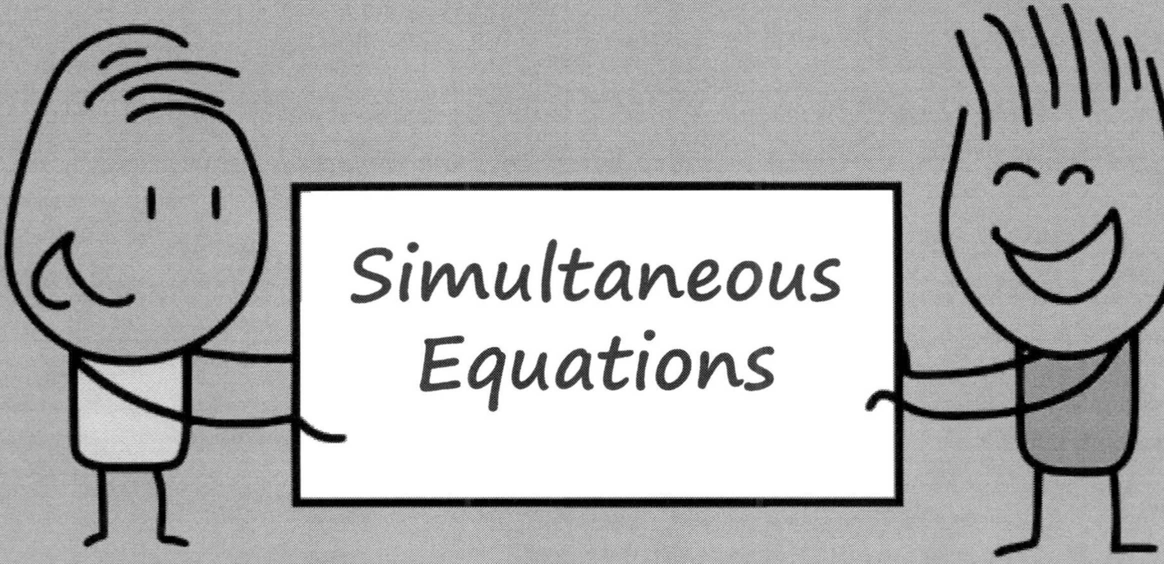
Simultaneous
Equations

# The Substitution Method

## Multiple Choice

**1** **Option B**

$y = x + 5$ (1)
$3x + y = 13$ (2)

Sub (1) into (2)

$3x + (x + 5) = 13$
$4x + 5 = 13$
$4x = 8$
$x = 2$

**2** **Option C**

$y = x + 3$ (1)
$6x + y = 17$ (2)

$x = y - 3$ (3)

Sub (3) into (2)

$6x + y = 17$
$6(y - 3) + y = 17$
$6y - 18 + y = 17$
$7y - 18 = 17$
$7y = 35$
$y = 5$

**3** **Option C**

$x = y - 1$ (1)
$2x + 3y = 8$ (2)

$y = x + 1$ (3)

Sub (3) into (2)

$2x + 3y = 8$
$2x + 3(x + 1) = 8$
$2x + 3x + 3 = 8$
$5x + 3 = 8$
$5x = 5$
$x = 1$

**4** **Option B**

$4a + 5b = 6$ (1)
$a - 2b = -5$ (2)

$a = 2b - 5$ (3)

Sub (3) into (1)

$4a + 5b = 6$
$4(2b - 5) + 5b = 6$
$8b - 20 + 5b = 6$
$13b - 20 = 6$
$13b = 26$
$b = 2$

Sub $b = 2$ into (3)

$a = 2b - 5$

$a = 2(2) - 5$

$a = 4 - 5$

$a = -1$

**5** **Option D**

$3m + n = 5$ (1)
$5m - 4n = -3$ (2)

$n = 5 - 3m$ (3)

Sub (3) into (2)

$5m - 4n = -3$
$5m - 4(5 - 3m) = -3$

$5m - 20 + 12m = -3$

$17m - 20 = -3$

$17m = 17$
$m = 1$

Sub $m = 1$ into (3)

$n = 5 - 3m$
$n = 5 - 3(1)$
$n = 5 - 3$
$n = 2$

## Time To Practice

**1** **a**

$y = x$ (1)
$4x + 3y = 7$ (2)

Sub (1) into (2)

$4x + 3y = 7$
$4x + 3x = 7$
$7x = 7$
$x = 1$

Sub $x = 1$ into (1)

$y = x$
$y = 1$

$\therefore x = 1\ and\ y = 1$

**b**

$y = x$ (1)
$7x + 3y = 20$ (2)

Sub (1) into (2)

$7x + 3y = 20$
$7x + 3x = 20$
$10x = 20$
$x = 2$

Sub $x = 2$ into (1)

$y = x$
$y = 2$

$\therefore x = 2\ and\ y = 2$

**c**

$y = x - 1$ (1)
$3x + 2y = 8$ (2)

Sub (1) into (2)

$3x + 2y = 8$
$3x + 2(x - 1) = 8$
$3x + 2x - 2 = 8$
$5x - 2 = 8$
$5x = 10$
$x = 2$

Sub $x = 2$ into (1)

$y = x - 1$
$y = 2 - 1$
$y = 1$

$\therefore x = 2\ and\ y = 1$

**d**

$y = x + 1$ (1)
$3x - 2y = -5$ (2)

Sub (1) into (2)

$3x - 2y = -5$
$3x - 2(x + 1) = -5$
$3x - 2x - 2 = -5$
$x - 2 = -5$
$x = -3$

Sub $x = -3$ into (1)

$y = x + 1$
$y = -3 + 1$
$y = -2$

$\therefore x = -3\ and\ y = -2$

**1** **e**

$x = 2y + 1$ (1)
$11y - 5x = -2$ (2)

Sub (1) into (2)

$11y - 5x = -2$
$11y - 5(2y + 1) = -2$
$11y - 10y - 5 = -2$
$y - 5 = -2$
$y = 3$

Sub $y = 1$ into (1)

$x = 2y + 1$
$x = 2(3) + 1$
$x = 7$

$\therefore x = 7 \; and \; y = 3$

**f**

$x = 3y - 3$ (1)
$7y - 2x = 11$ (2)

Sub (1) into (2)

$7y - 2x = 11$
$7y - 2(3y - 3) = 11$
$7y - 6y + 6 = 11$
$y + 6 = 11$
$y = 5$

Sub $y = 5$ into (1)

$x = 3y - 3$
$x = 3(5) - 3$
$x = 12$

$\therefore x = 12 \; and \; y = 5$

**g**

$x = 2y + 12$ (1)
$3x + 2y = -4$ (2)

Sub (1) into (2)

$3x + 2y = -4$
$3(2y + 12) + 2y = -4$
$6y + 36 + 2y = -4$
$8y + 36 = -4$
$8y = -40$
$y = -5$

Sub $y = -5$ into (1)

$x = 2y + 12$
$x = 2(-5) + 12$
$x = -10 + 12$
$x = 2$

$\therefore x = 2 \; and \; y = -5$

**h**

$x = 4y + 2$ (1)
$2y - 3x = 14$ (2)

Sub (1) into (2)

$2y - 3x = 14$
$2y - 3(4y + 2) = 14$
$2y - 12y - 6 = 14$
$-10y - 6 = 14$
$-10y = 20$
$y = -2$

Sub $y = -2$ into (1)

$x = 4y + 2$
$x = 4(-2) + 2$
$x = -8 + 2$
$x = -6$

$\therefore x = -6 \; and \; y = -2$

**2** **a**

$3x + 4y = 27$ (1)
$2x - y = -4$ (2)

Rearrange (2)

$2x - y = -4$
$y = 2x + 4$ (3)

Sub (3) into (1)

$3x + 4y = 27$
$3x + 4(2x + 4) = 27$
$3x + 8x + 16 = 27$
$11x + 16 = 27$
$11x = 11$
$x = 1$

Sub $x = 1$ into (3)

$y = 2x + 4$
$y = 2(1) + 4$
$y = 6$

$\therefore x = 1 \ and \ y = 6$

**b**

$3x - y = 23$ (1)
$2x - 2y = 10$ (2)

Rearrange (1)

$3x - y = 23$
$y = 3x - 23$ (3)

Sub (3) into (2)

$2x - 2y = 10$
$2x - 2(3x - 23) = 10$
$2x - 6x + 46 = 10$
$-4x + 46 = 10$
$-4x = -36$
$x = 9$

Sub $x = 9$ into (3)

$y = 3x - 23$
$y = 3(9) - 23$
$y = 4$

$\therefore x = 9 \ and \ y = 4$

**2** **c** $4a + 3b = 23$ (1)
$2a - b = -1$ (2)

Rearrange (2)

$2a - b = -1$
$b = 2a + 1$ (3)

Sub (3) into (1)

$4a + 3b = 23$
$4a + 3(2a + 1) = 23$
$4a + 6a + 3 = 23$
$10a + 3 = 23$
$10a = 20$
$a = 2$

Sub $a = 2$ into (3)

$b = 2a + 1$
$b = 2(2) + 1$
$b = 5$

$\therefore a = 2 \text{ and } b = 5$

**d** $2a + b = 16$ (1)
$3a - 4b = 24$ (2)

Rearrange (1)

$2a + b = 16$
$b = 16 - 2a$ (3)

Sub (3) into (2)

$3a - 4b = 24$
$3a - 4(16 - 2a) = 24$
$3a - 64 + 8a = 24$
$11a - 64 = 24$
$11a = 88$
$a = 8$

Sub $a = 8$ into (3)

$b = 16 - 2a$
$b = 16 - 2(8)$
$b = 0$

$\therefore x = 8 \text{ and } y = 0$

**2** **e**

$5p - q = -22$ (1)
$7p + 2q = -7$ (2)

Rearrange (1)

$5p - q = -22$
$q = 5p + 22$ (3)

Sub (3) into (2)

$7p + 2q = -7$
$7p + 2(5p + 22) = -7$
$7p + 10p + 44 = -7$
$17p + 44 = -7$
$17p = -51$
$p = -3$

Sub $p = -3$ into (3)

$q = 5p + 22$
$q = 5(-3) + 22$
$q = -15 + 22$
$q = 7$

$\therefore p = -3 \text{ and } q = 7$

**f**

$5p + 2q = 5$ (1)
$3p - q = 25$ (2)

Rearrange (2)

$3p - q = 25$
$q = 3p - 25$ (3)

Sub (3) into (1)

$5p + 2q = 5$
$5p + 2(3p - 25) = 5$
$5p + 6p - 50 = 5$
$11p - 50 = 5$
$11p = 55$
$p = 5$

Sub $p = 5$ into (3)

$q = 3p - 25$
$q = 3(5) - 25$
$q = 15 - 25$
$q = -10$

$\therefore p = 5 \text{ and } q = -10$

**2** **g**

$3m - 2n = 34$ (1)
$m + 4n = 16$ (2)

Rearrange (2)

$m + 4n = 16$
$m = 16 - 4n$ (3)

Sub (3) into (1)

$3m - 2n = 34$
$3(16 - 4n) - 2n = 34$
$48 - 12n - 2n = 34$
$48 - 14n = 34$
$48 - 34 = 14n$
$14 = 14n$
$n = 1$

Sub $n = 1$ into (3)

$m = 16 - 4n$
$m = 16 - 4(1)$
$m = 12$

$\therefore m = 12 \ and \ n = 1$

**h**

$6m - 3n = 21$ (1)
$2m + n = -15$ (2)

Rearrange (2)

$2m + n = -15$
$n = -15 - 2m$ (3)

Sub (3) into (1)

$6m - 3n = 21$
$6m - 3(-15 - 2m) = 21$
$6m + 45 + 6m = 21$
$12m + 45 = 21$
$12m = -24$
$m = -2$

Sub $m = -2$ into (3)

$n = -15 - 2m$
$n = -15 - 2(-2)$
$n = -15 + 4$
$n = -11$

$\therefore m = -2 \ and \ n = -11$

**2** **i**

$5r + 3s = 44$ (1)
$2r - s = 0$ (2)

Rearrange (2)

$2r - s = 0$
$s = 2r$ (3)

Sub (3) into (1)

$5r + 3s = 44$
$5r + 3(2r) = 44$
$5r + 6r = 44$
$11r = 44$
$r = 4$

Sub $r = 4$ into (3)

$s = 2r$
$s = 2(4)$
$s = 8$

$\therefore r = 4 \ and \ s = 8$

**j**

$4r + 3s = -59$ (1)
$r - 2s = 10$ (2)

Rearrange (2)

$r - 2s = 10$
$r = 2s + 10$ (3)

Sub (3) into (1)

$4r + 3s = -59$
$4(2s + 10) + 3s = -59$
$8s + 40 + 3s = -59$
$11s + 40 = -59$
$11s = -99$
$s = -9$

Sub $s = -9$ into (3)

$r = 2s + 10$
$r = 2(-9) + 10$
$r = -18 + 10$
$r = -8$

$\therefore r = -8 \ and \ s = -9$

## Challenge Yourself

**1** **a**

$m - 2n - 5 = 0$ (1)
$3m - n - 25 = 0$ (2)

Rearrange (1)

$m - 2n - 5 = 0$
$m = 2n + 5$ (3)

Sub (3) into (2)

$3m - n - 25 = 0$
$3(2n + 5) - n - 25 = 0$
$6n + 15 - n - 25 = 0$
$5n - 10 = 0$
$5n = 10$
$n = 2$

Sub $n = 2$ into (3)

$m = 2n + 5$
$m = 2(2) + 5$
$m = 9$

$\therefore m = 9 \text{ and } n = 2$

**b**

$y + x - 10 = 0$ (1)
$7y + 2x - 5 = 0$ (2)

Rearrange (1)

$y + x - 10 = 0$
$x = 10 - y$ (3)

Sub (3) into (2)

$7y + 2x - 5 = 0$
$7y + 2(10 - y) - 5 = 0$
$7y + 20 - 2y - 5 = 0$
$5y + 15 = 0$
$5y = -15$
$y = -3$

Sub $y = -3$ into (3)

$x = 10 - y$
$x = 10 - (-3)$
$x = 13$

$\therefore x = 13 \text{ and } y = -3$

**1** **c**

$r - 2s = 4$ (1)

$r + \frac{s}{3} = 4$ (2)

Rearrange (1)

$r - 2s = 4$

$r = 2s + 4$ (3)

Sub (3) into (2)

$r + \frac{s}{3} = 4$

$2s + 4 + \frac{s}{3} = 4$

$2s + \frac{s}{3} = 0$

$\frac{2s}{1} + \frac{s}{3} = 0$

$\frac{6s + 3s}{3} = 0$

$9s = 0$

$s = 0$

Sub $s = 0$ into (3)

$r = 2s + 4$

$r = 2(0) + 4$

$r = 4$

$\therefore r = 4 \text{ and } s = 0$

**d**

$2c + \frac{d}{4} = 7$ (1)

$c - \frac{d}{2} = 1$ (2)

Rearrange (2)

$c - \frac{d}{2} = 1$

$c = \frac{d}{2} + 1$ (3)

Sub (3) into (1)

$2c + \frac{d}{4} = 7$

$2\left(\frac{d}{2} + 1\right) + \frac{d}{4} = 7$

$d + 2 + \frac{d}{4} = 7$

$d + \frac{d}{4} = 5$

$\frac{d}{1} + \frac{d}{4} = 5$

$\frac{4d + d}{4} = 5$

$5d = 20$

$d = 4$

Sub $d = 4$ into (3)

$c = \frac{d}{2} + 1$

$c = \frac{4}{2} + 1 = 3$

$\therefore c = 3 \text{ and } d = 4$

**1** **e**

$3p + q = 5$ (1)

$p - \frac{q}{3} = 1$ (2)

Rearrange (1)

$3p + q = 5$

$q = 5 - 3p$ (3)

Sub (3) into (2)

$p - \frac{q}{3} = 1$

$p - \frac{5 - 3p}{3} = 1$

$\frac{p}{1} - \frac{5 - 3p}{3} = 1$

$\frac{3p - (5 - 3p)}{3} = 1$

$\frac{6p - 5}{3} = 1$

$6p - 5 = 3$

$6p = 8$

$p = \frac{8}{6} = \frac{4}{3}$

Sub $\boldsymbol{p} = \frac{4}{3}$ into (3)

$q = 5 - 3p$

$q = 5 - 3\left(\frac{4}{3}\right)$

$q = 5 - 4$

$q = 1$

$\therefore p = \frac{4}{3}$ $and$ $q = 1$

**f**

$2a - b = 3$ (1)

$a + \frac{b}{2} = 2$ (2)

Rearrange (1)

$2a - b = 3$

$b = 2a - 3$ (3)

Sub (3) into (2)

$a + \frac{b}{2} = 2$

$a + \frac{2a - 3}{2} = 2$

$\frac{2a}{2} + \frac{2a - 3}{2} = 2$

$\frac{4a - 3}{2} = 2$

$4a - 3 = 4$

$4a = 7$

$a = \frac{7}{4}$

Sub $\boldsymbol{a} = \frac{7}{4}$ into (3)

$b = 2a - 3$

$b = 2\left(\frac{7}{4}\right) - 3$

$b = \frac{7}{2} - \frac{3}{1} = \frac{7 - 6}{2}$

$b = \frac{1}{2}$

$\therefore c = \frac{7}{4}$ $and$ $d = \frac{1}{2}$

# The Elimination Method

## Multiple Choice

**1** **Option A**

$4a - 3b = -10$ (1)
$2a + 3b = 4$ (2)

$(\mathbf{1}) + (\mathbf{2})$

$4a + 2a = -10 + 4$
$6a = -6$

**2** **Option D**

$-5x + 2y = 16$ (1)
$-5x - 3y = 1$ (2)

$(\mathbf{1}) - (\mathbf{2})$

$2y - (-3y) = 16 - 1$
$5y = 15$

**3** **Option A**

$3x + 4y = 2$ (1)
$5x + 2y = -6$ (2)

$\mathbf{2} \times (\mathbf{2})$

$2 \times (5x + 2y = -6)$ (3)
$10x + 4y = -12$

$3x + 4y = 2$ (1)
$10x + 4y = -12$ (3)

$(\mathbf{3}) - (\mathbf{1})$

$10x - 3x = -12 - 2$
$7x = -14$
$x = -2$

**4** **Option B**

$-2x + 3y = 7$ (1)
$3x - 2y = -3$ (2)

$\mathbf{3} \times (\mathbf{1})$ *and* $\mathbf{2} \times (\mathbf{2})$

$3 \times (-2x + 3y = 7)$
$2 \times (3x - 2y = -3)$

$-6x + 9y = 21$ (3)
$6x - 4y = -6$ (4)

$(\mathbf{4}) + (\mathbf{3})$

$-4y + 9y = -6 + 21$
$5y = 15$
$y = 3$

**5** **Option C**

$3x + 4y = 13$ (1)
$2x - 5y = 1$ (2)

$\mathbf{2} \times (\mathbf{1})$ *and* $\mathbf{3} \times (\mathbf{2})$

$2 \times (3x + 4y = 13)$
$3 \times (2x - 5y = 1\,)$

$6x + 8y = 26$ (3)
$6x - 15y = 3$ (4)

$(\mathbf{3}) - (\mathbf{4})$

$8y - (-15y) = 26 - 3$
$23y = 23$
$y = 1$

**Sub** $y = \mathbf{1}$ **into (2)**

$2x - 5y = 1$
$2x - 5(1) = 1$
$2x - 5 = 1$
$2x = 6$
$x = 3$

$\therefore x = 3 \text{ and } y = 1$

## Time To Practice

**1** **a**

$3x + 5y = 21 \quad (1)$
$2x + 5y = 19 \quad (2)$

$(1) - (2)$

$3x - 2x = 21 - 19$
$x = 2$

Sub $x = 2$ into $(2)$

$2x + 5y = 19$
$2(2) + 5y = 19$
$4 + 5y = 19$
$5y = 15$
$y = 3$

$\therefore x = 2 \text{ and } y = 3$

**b**

$3x + 5y = 17 \quad (1)$
$3x - 4y = -19 \quad (2)$

$(1) - (2)$

$5y - (-4y) = 17 - (-19)$
$5y + 4y = 17 + 19$
$9y = 36$
$y = 4$

Sub $y = 4$ into $(1)$

$3x + 5y = 17$
$3x + 5(4) = 17$
$3x + 20 = 17$
$3x = -3$
$x = -1$

$\therefore x = -1 \text{ and } y = 4$

**c**

$6a - 4b = 14 \quad (1)$
$-2a - 4b = -10 \quad (2)$

$(1) - (2)$

$6a - (-2a) = 14 - (-10)$
$8a = 24$
$a = 3$

Sub $a = 3$ into $(1)$

$6a - 4b = 14$
$6(3) - 4b = 14$
$18 - 4b = 14$
$-4b = -4$
$b = 1$

$\therefore a = 3 \text{ and } b = 1$

**d**

$7a + 6b = -8 \quad (1)$
$4a - 6b = -14 \quad (2)$

$(1) + (2)$

$7a + 4a = -8 + (-14)$
$11a = -22$
$a = -2$

Sub $a = -2$ into $(2)$

$4a - 6b = -14$
$4(-2) - 6b = -14$
$-8 - 6b = -14$
$-6b = -6$
$b = 1$

$\therefore a = -2 \text{ and } b = 1$

**1** **e**

$11p - 4q = 2$ (1)
$7p - 4q = -6$ (2)

$(\mathbf{1}) - (\mathbf{2})$

$11p - 7p = 2 - (-6)$
$4p = 8$
$p = 2$

**Sub $p = 2$ into (1)**

$11p - 4q = 2$
$11(2) - 4q = 2$
$22 - 4q = 2$
$20 = 4q$
$q = 5$

$\therefore p = 2 \; and \; q = 5$

**f**

$3p - 5q = 6$ (1)
$-2p - 5q = 46$ (2)

$(\mathbf{2}) - (\mathbf{1})$

$-2p - 3p = 46 - 6$
$-5p = 40$
$p = -8$

**Sub $p = -8$ into (1)**

$3p - 5q = 6$
$3(-8) - 5q = 6$
$-24 - 5q = 6$
$-5q = 30$
$q = -6$

$\therefore p = -8 \; and \; q = -6$

**g**

$7x - 3y = -16$ (1)
$-7x + 2y = 6$ (2)

$(\mathbf{1}) + (\mathbf{2})$

$-3y + 2y = -16 + 6$
$-y = 10$
$y = 10$

**Sub $y = 10$ into (2)**

$-7x + 2y = 6$
$-7x + 2(10) = 6$
$-7x + 20 = 6$
$-7x = -14$
$x = 2$

$\therefore x = 2 \; and \; y = 10$

**h**

$-3x + 4y = 14$ (1)
$5x + 4y = -34$ (2)

$(\mathbf{1}) - (\mathbf{2})$

$-3x - 5x = 14 - (-34)$
$-8x = 18 + 34$
$-8x = 48$
$x = -6$

**Sub $x = -6$ into (1)**

$-3x + 4y = 14$
$-3(-6) + 4y = 14$
$18 + 4y = 14$
$4y = -4$
$y = -1$

$\therefore x = -6 \; and \; y = -1$

**2** **a**

$2x + 3y = 9$ (1)
$4x + 5y = 15$ (2)

$2 \times (1)$

$2 \times (2x + 3y = 9)$
$4x + 6y = 18$ (3)

$4x + 5y = 15$ (2)
$4x + 6y = 18$ (3)

$(3) - (2)$

$6y - 5y = 18 - 15$
$y = 3$

Sub $y = 3$ into (1)

$2x + 3y = 9$
$2x + 3(3) = 9$
$2x + 9 = 9$
$2x = 0$
$x = 0$

$\therefore x = 0 \text{ and } y = 3$

**b**

$3x + 7y = 1$ (1)
$9x - 4y = -22$ (2)

$3 \times (1)$

$3 \times (3x + 7y = 1)$
$9x + 21y = 3$ (3)

$9x - 4y = -22$ (2)
$9x + 21y = 3$ (3)

$(3) - (2)$

$21y - (-4y) = 3 - (-22)$
$21y + 4y = 3 + 22$
$25y = 25$
$y = 1$

Sub $y = 1$ into (1)

$3x + 7y = 1$
$3x + 7(1) = 1$
$3x + 7 = 1$
$3x = -6$
$x = -2$

$\therefore x = -2 \text{ and } y = 1$

**2** **c** $12a + 3b = 3$ (1)
$-4a - 5b = 11$ (2)

$3 \times (2)$

$3 \times (-4a - 5b = 11)$
$-12a - 15b = 33$ (3)

$12a + 3b = 3$ (1)
$-12a - 15b = 33$ (3)

$(3) + (1)$

$-15b + 3b = 33 + 3$
$-12b = 36$
$b = -3$

Sub $b = -3$ into (1)

$12a + 3b = 3$
$12a + 3(-3) = 3$
$12a - 9 = 3$
$12a = 12$
$a = 1$

$\therefore a = 1 \text{ and } b = -3$

**d** $7a - 6b = -9$ (1)
$2a + 12b = -30$ (2)

$2 \times (1)$

$2 \times (7a - 6b = -9)$
$14a - 12b = -18$ (3)

$2a + 12b = -30$ (2)
$14a - 12b = -18$ (3)

$(3) + (2)$

$14a + 2a = -18 + (-30)$
$16a = -18 - 30$
$16a = -48$
$a = -3$

Sub $a = -3$ into (1)

$7a - 6b = -9$
$7(-3) - 6b = -9$
$-21 - 6b = -9$
$-6b = 12$
$b = -2$

$\therefore a = -3 \text{ and } b = -2$

**2** **e**

$5p + 2q = -7$ (1)
$2p - 20q = 18$ (2)

$10 \times (1)$

$10 \times (5p + 2q = -7)$
$50p + 20q = -70$ (3)

$2p - 20q = 18$ (2)
$50p + 20q = -70$ (3)

$(2) + (3)$

$2p + 50p = 18 + (-70)$
$52p = -52$
$p = -1$

Sub $p = -1$ into (1)

$5p + 2q = -7$
$5(-1) + 2q = -7$
$-5 + 2q = -7$
$2q = -2$
$q = -1$

$\therefore p = -1 \text{ and } q = -1$

**f**

$2p - 3q = 0$ (1)
$5p + 15q = 45$ (2)

$5 \times (1)$

$5 \times (2p - 3q = 0)$
$10p - 15q = 0$ (3)

$5p + 15q = 45$ (2)
$10p - 15q = 0$ (3)

$(3) + (2)$

$10p + 5p = 45$
$15p = 45$
$p = 3$

Sub $p = 3$ into (1)

$2p - 3q = 0$
$2(3) - 3q = 0$
$6 - 3q = 0$
$6 = 3q$
$q = 2$

$\therefore p = 3 \text{ and } q = 2$

**2** **g** $-2x + 5y = -10$ (1)

$8x - 3y = 6$ (2)

$4 \times (1)$

$4 \times (-2x + 5y = -10)$

$-8x + 20y = -40$ (3)

$8x - 3y = 6$ (2)

$-8x + 20y = -40$ (3)

$(3) + (2)$

$20y + (-3y) = -40 + 6$

$20y - 3y = -34$

$20y - 3y = -34$

$17y = -34$

$y = -2$

Sub $y = -2$ into (2)

$8x - 3y = 6$

$8x - 3(-2) = 6$

$8x + 6 = 6$

$8x = 0$

$x = 0$

$\therefore x = 0 \; and \; y = -2$

**h** $-5x - 6y = 4$ (1)

$3x - 2y = 20$ (2)

$3 \times (2)$

$3 \times (3x - 2y = 20)$

$9x - 6y = 60$ (3)

$-5x - 6y = 4$ (1)

$9x - 6y = 60$ (3)

$(3) - (1)$

$9x - (-5x) = 60 - 4$

$9x + 5x = 56$

$14x = 56$

$x = 4$

Sub $x = 4$ into (2)

$3x - 2y = 20$

$3(4) - 2y = 20$

$12 - 2y = 20$

$-2y = 8$

$y = -4$

$\therefore x = 4 \; and \; y = -4$

**3** **a**

$3x + 2y = 7$ (1)
$2x + 5y = 12$ (2)

$2 \times (1)$
$3 \times (2)$

$2 \times (3x + 2y = 7)$
$3 \times (2x + 5y = 12\ )$

$6x + 4y = 14$ (3)
$6x + 15y = 36$ (4)

$(4) - (3)$

$15y - 4y = 36 - 14$
$11y = 22$
$y = 2$

Sub $y = 2$ into (2)

$2x + 5y = 12$
$2x + 5(2) = 12$
$2x + 10 = 12$
$2x = 2$
$x = 1$

$\therefore x = 1\ and\ y = 2$

**b**

$-3x + 2y = -7$ (1)
$4x - 5y = 7$ (2)

$5 \times (1)$
$2 \times (2)$

$5 \times (-3x + 2y = -7)$
$2 \times (4x - 5y = 7)$

$-15x + 10y = -35$ (3)
$8x - 10y = 14$ (4)

$(4) + (3)$

$8x + (-15x) = 14 + (-35)$
$8x - 15x = 14 - 35$
$-7x = -21$
$x = 3$

Sub $x = 3$ into (2)

$4x - 5y = 7$
$4(3) - 5y = 7$
$12 - 5y = 7$
$12 - 7 = 5y$
$5y = 5$
$y = 1$

$\therefore x = 3\ and\ y = 1$

**3** **c**

$4a + 3b = -5$ (1)
$3a - 2b = 9$ (2)

$2 \times (1)$
$3 \times (2)$

$2 \times (4a + 3b = -5)$
$3 \times (3a - 2b = 9)$

$8a + 6b = -10$ (3)
$9a - 6b = 27$ (4)

$(4) + (3)$

$9a + 8a = 27 + (-10)$
$17a = 17$
$a = 1$

Sub $a = 1$ into (2)

$3a - 2b = 9$
$3(1) - 2b = 9$
$3 - 2b = 9$
$-2b = 6$
$b = -3$

$\therefore a = 1 \text{ and } b = -3$

**d**

$-3a + 4b = -2$ (1)
$-4a - 5b = 18$ (2)

$4 \times (1)$
$3 \times (2)$

$4 \times (-3a + 4b = -2)$
$3 \times (-4a - 5b = 18)$

$-12a + 16b = -8$ (3)
$-12a - 15b = 54$ (4)

$(3) - (4)$

$16b - (-15b) = -8 - 54$
$16b + 15b = -62$
$31b = -62$
$b = -2$

Sub $b = -2$ into (1)

$-3a + 4b = -2$
$-3a + 4(-2) = -2$
$-3a - 8 = -2$
$-3a = 6$
$a = -2$

$\therefore a = -2 \text{ and } b = -2$

**3** **e**

$7p - 3q = 5$ (1)
$-5p + 4q = -11$ (2)

$4 \times (1)$
$3 \times (2)$

$4 \times (7p - 3q = 5)$
$3 \times (-5p + 4q = -11)$

$28p - 12q = 20$ (3)
$-15p + 12q = -33$ (4)

$(4) + (3)$

$-15p + 28p = -33 + 20$
$13p = -13$
$p = -1$

Sub $p = -1$ into (1)

$7p - 3q = 5$
$7(-1) - 3q = 5$
$-7 - 3q = 5$
$-3q = 12$
$q = -4$

$\therefore p = -1 \text{ and } q = -4$

**f**

$-2p + 5q = 2$ (1)
$9p - 3q = 30$ (2)

$9 \times (1)$
$2 \times (2)$

$9 \times (-2p + 5q = 2)$
$2 \times (9p - 3q = 30)$

$-18p + 45q = 18$ (3)
$18p - 6q = 60$ (4)

$(3) + (4)$

$45q + (-6q) = 18 + 60$
$45q - 6q = 78$
$39q = 78$
$q = 2$

Sub $q = 2$ into (1)

$9p - 3q = 30$
$9p - 3(2) = 30$
$9p - 6 = 30$
$9p = 36$
$p = 4$

$\therefore p = 4 \text{ and } q = 2$

**3** **g**

$-8x - 3y = 7$ (1)
$5x + 2y = -5$ (2)

$2 \times (1)$
$3 \times (2)$

$2 \times (-8x - 3y = 7)$
$3 \times (5x + 2y = -5)$

$-16x - 6y = 14$ (3)
$15x + 6y = -15$ (4)

$(3) + (4)$

$-16x + 15x = 14 + (-15)$
$-x = 14 - 15$
$-x = -1$
$x = 1$

Sub $x = 1$ into (2)

$5x + 2y = -5$
$5(1) + 2y = -5$
$5 + 2y = -5$
$2y = -10$
$y = -5$

$\therefore x = 1 \text{ and } y = -5$

**h**

$-3x - 7y = -9$ (1)
$-5x - 4y = 8$ (2)

$5 \times (1)$
$3 \times (2)$

$5 \times (-3x - 7y = -9)$
$3 \times (-5x - 4y = 8)$

$-15x - 35y = -45$ (3)
$-15x - 12y = 24$ (4)

$(4) - (3)$

$-12y + 35y = 24 + 45$
$23y = 69$
$y = 3$

Sub $y = 3$ into (2)

$-5x - 4y = 8$
$-5x - 4(3) = 8$
$-5x - 12 = 8$
$-5x = 20$
$x = -4$

$\therefore x = -4 \text{ and } y = 3$

## Challenge Yourself

**1** **a**

$3 + 4x - 3y = 0$
$4 - 2y + 3x = 0$

$4x - 3y = -3$ (1)
$3x - 2y = -4$ (2)

$2 \times (1)$
$3 \times (2)$

$2 \times (4x - 3y = -3)$
$3 \times (3x - 2y = -4)$

$8x - 6y = -6$ (3)
$9x - 6y = -12$ (4)

$(4) - (3)$

$9x - 8x = -12 - (-6)$
$x = -12 + 6$
$x = -6$

Sub $x = -6$ into (2)

$3x - 2y = -4$
$3(-6) - 2y = -4$
$-18 - 2y = -4$
$-2y = 14$
$y = -7$

$\therefore x = -6 \text{ and } y = -7$

**b**

$2x = 15 + 5y$
$-7y = 6x - 23$

$2x - 5y = 15$ (1)
$-6x - 7y = -23$ (2)

$3 \times (1)$

$3 \times (2x - 5y = 15)$ (3)
$6x - 15y = 45$

$-6x - 7y = -23$ (2)
$6x - 15y = 45$ (3)

$(2) + (3)$

$-7y - 15y = -23 + 45$
$-22y = 22$
$y = -1$

Sub $y = -1$ into (1)

$2x - 5y = 15$
$2x - 5(-1) = 15$
$2x + 5 = 15$
$2x = 10$
$x = 5$

$\therefore x = 5 \text{ and } y = -1$

**1** **c**

$$\frac{2m - 5n}{2} = 16 \qquad (1)$$

$$2m + 3n = 0 \qquad (2)$$

$Get\ rid\ of\ fraction\ (1)$

$$\frac{2m - 5n}{2} = 16$$

$$2m - 5n = 32 \qquad (3)$$

$$2m + 3n = 0 \qquad (2)$$
$$2m - 5n = 32 \qquad (3)$$

$(2) - (3)$

$3n - (-5n) = 0 - 32$
$3n + 5n = -32$
$8n = -32$
$n = -4$

Sub $n = -4$ into (2)

$2m + 3n = 0$

$2m + 3(-4) = 0$

$2m - 12 = 0$

$2m = 12$
$m = 6$

$\therefore m = 6\ and\ n = -4$

**d**

$$\frac{2p}{3} + 3q = -4 \qquad (1)$$

$$5p - 9q = 33 \qquad (2)$$

$3 \times (1)$

$$3 \times \left(\frac{2p}{3} + 3q = -4\right)$$

$$2p + 9q = -12 \qquad (3)$$

$$5p - 9q = 33 \qquad (2)$$
$$2p + 9q = -12 \qquad (3)$$

$(2) + (3)$

$5p + 2p = 33 + (-12)$
$7p = 33 - 12$
$7p = 21$
$p = 3$

Sub $p = 3$ into (2)

$5p - 9q = 33$

$5(3) - 9q = 33$

$15 - 9q = 33$
$-9q = 18$
$q = -2$

$\therefore p = 3\ and\ q = -2$

**1** **e**

$$\frac{x+2}{2}+\frac{y-1}{3}=1 \qquad (1)$$

$$5x+2y=-6 \qquad (2)$$

*Get rid of fractions* (1)

$$\frac{x+2}{2}+\frac{y-1}{3}=1$$

$$\frac{3(x+2)+2(y-1)}{6}=1$$

$$\frac{3x+6+2y-2}{6}=1$$

$$\frac{3x+2y+4}{6}=1$$

$$3x+2y+4=6$$

$$3x+2y=2 \qquad (3)$$

$$5x+2y=-6 \qquad (2)$$
$$3x+2y=2 \qquad (3)$$

(2) − (3)

$$5x-3x=-6-3$$
$$2x=-8$$
$$x=-4$$

Sub $x=-4$ into (2)

$$5x+2y=-6$$
$$5(-4)+2y=-6$$
$$-20+2y=-6$$
$$2y=14$$
$$y=7$$

$\therefore x=-4 \text{ and } y=7$

**1** **f** $\frac{2x}{3}+\frac{y}{2}=\frac{1}{6}$ (1)

$\frac{x}{5}+\frac{3y}{2}=-\frac{2}{5}$ (2)

*Get rid of fractions* (1)

$\frac{2x}{3}+\frac{y}{2}=\frac{1}{6}$

$\frac{4x+3y}{6}=\frac{1}{6}$

$4x+3y=1$ (3)

*Get rid of fractions* (2)

$\frac{x}{5}+\frac{3y}{2}=-\frac{2}{5}$

$\frac{2x+15y}{10}=-\frac{2}{5}$

$2x+15y=-4$ (4)

$4x+3y=1$ (3)
$2x+15y=-4$ (4)

**2 × (4)**

$2\times(2x+15y=-4)$
$4x+30y=-8$ (5)

$4x+3y=1$ (3)
$4x+30y=-8$ (5)

(5) − (3)

$30y-3y=-8-1$
$27y=-9$

$y=-\frac{9}{27}=-\frac{1}{3}$

**Sub $y=-\frac{1}{3}$ into (3)**

$4x+3y=1$
$4x+3\left(-\frac{1}{3}\right)=1$

$4x-1=1$
$4x=2$

$x=\frac{2}{4}=\frac{1}{2}$

$\therefore x=\frac{1}{2}$ *and* $y=-\frac{1}{3}$

## Was the answer book any good?

I genuinely hope this answer book helped you learn.

But I won't know unless you leave a review.

So, if you don't mind, can you write one?

Tell me what you think.

Was this any good?

Were the steps clear?

Is your understanding better after using this book?

Any mistakes?

Better not be.

I triple checked everything.

But seriously, let me know.

Anything.

Good or bad.

Appreciate it.

Made in the USA
Coppell, TX
05 November 2026

11302305R00098